Europe and the New Technologies

Europe and the New Technologies

Six Case Studies in Innovation and Adjustment

Edited by
Margaret Sharp

Senior Research Fellow
Science Policy Research Unit
University of Sussex

Frances Pinter (Publishers), London

© Margaret Sharp 1983

First published in Great Britain in 1985 by
Frances Pinter (Publishers) Limited
25 Floral Street, London WC2E 9DS

British Library Cataloguing in Publication Data
Europe and the new technologies: six case
    studies in innovation and adjustment.
    1. Technological innovations—Europe   2. Europe
    —Industries
    I. Sharp, Margaret, *1938-*
    338'.06   T173.8

ISBN 0-86187-553-2

Typeset by Joshua Associates Ltd, Oxford
Printed by SRP Ltd, Exeter

Contents

Tables	vii
Figures	x
Contributors	xi
Notes on text	xiii
Preface	xv

1 Introduction — 1
Margaret Sharp

The concept of a new industrial activity	2
The choice of case studies	3
The case studies: framework	4
The case studies: a variety of interests and experiences	5
The case studies: an overview	8

2 Computer aided design: Europe's role and American technology — 10
Erik Arnold and Peter Senker

CAD technology and competition	11
Government policies in individual countries	19
Policy implications	37
The case for a European initiative	41

3 Advanced machine tools: production, diffusion and trade — 46
Ernst-Jürgen Horn, Henning Klodt and Christopher Saunders

Recent developments in machine tools	47
Diffusion patterns	55
Country-based developments	61
An analysis of EEC foreign trade in advanced machine tools	68
Conclusions	78

4 Telecommunications: a challenge to the old order — 87
Godefroy Dang Nguyen

The background to present developments	89
The impact of electronics	99

	From telecommunications to videocommunications —the revolution in network provision	109
	The future: can the PTTs survive?	122
	A European scenario	127
5	**Videotex: much ado about nothing?** *Godefroy Dang Nguyen and Erik Arnold*	134
	The concept of the *filière*	135
	Videotex as a *microfilière*	138
	Videotex in practice: experience in the United Kingdom, West Germany and France	141
	Lessons and conclusions	154
6	**Biotechnology: watching and waiting** *Margaret Sharp*	161
	What is biotechnology?	162
	The industrial impact of biotechnology	166
	Country-based developments	174
	The European perspective	198
	Firms, governments and technologies	201
7	**The offshore supplies industry: fast, continuous and incremental change** *P. Lesley Cook*	213
	Newness is a matter of degree	213
	The main characteristics of the new activity	215
	Country-based experience	220
	Assessment of the policies	245
	Conclusions	253
	Glossary	261
8	**Conclusions: Technology gap or management gap?** *Margaret Sharp*	263
	The six case studies	263
	Some common themes	278
	Europe's capacity to compete	288
	How can Europe improve its performance?	292
Index		298

Tables

2.1	World-wide turnkey CAD market shares	16
2.2	Global sales of CAD/CAM turnkey systems by major United States vendors ($000s) 1969–83	17
2.3	French CAD market, estimates of shares of installed stock, 1982 (excluding IBM)	21
2.4	Estimates of West German CAD park, 1981, shares of suppliers (excluding IBM and Calma)	24
2.5	British turnkey CAD park, market shares, 1981	29
2.6	CAD/CAM—number and type of users in Sweden, 1982	33
2.7	International comparisons of CAD use, 1982	38
3.1	Shares in world machine tool production and exports	50
3.2	Shares in world exports of machine tools, 1965–81	50
3.3	Size of machine tool firms (1979)	51
3.4	Industrial origin of major producers of industrial robots	53
3.5	Stock of NC machine tools 1969 to 1980–1	57
3.6	Production of final consumption of NC machine tools 1976 and 1980	57
3.7	NC machine tools as percentage total machine tool park	58
3.8	Production of industrial robots 1974–85	59
3.9	Main producers of industrial robots 1981	60
3.10	Industrial robots installed 1974–85	61
3.11	NC machine tools: EEC exports 1976–83	69
3.12	NC machine tools: unit value of exports compared to total EEC exports 1976 and 1983	71
3.13	NC machine tools: unit values of exports in ECUs and national currencies; 1983 as % of 1976	72
3.14	NC machine tools: composition of exports compared with shares of total EEC exports and export unit values	74
3.15	NC machine tools: EEC imports from non-EEC countries 1983	77

3.16	NC machine tools: imports by principal EEC exporting countries 1983	78
4.1	Telephone penetration rates by country	90
4.2	Regional markets for telecommunications equipment	92
4.3	Telecommunications equipment manufacturing—the world market by type of equipment	93
4.4	The twelve largest manufacturers of telecommunications equipment ranked by sales in 1982	93
4.5	Market share of ITT, Ericsson and Siemens in the world open market	95
4.6	Switching market in Europe, 1974–1984	97
4.7	British Telecom—exchanges in use—1984	103
4.8	R&D cost of digital switching systems	108
4.9	Data transmission terminals in Europe, 1981	110
6.1	The industrial applications of biotechnology: summary table	167
6.2	Emergence of new biotechnology firms, 1977–1983	176
6.3	Proportion of firms in the United States pursuing applications of biotechnology in specific industrial sectors, 1982/3	178
6.4	R&D budgets for biotechnology from some leading American companies, 1982	179
6.5	American biotechnology patent grants, January 1963–December 1982	182
6.6	Government spending in biotechnology programmes—European governments compared to the United States and Japan	183
6.7	Publicly funded research spending in areas broadly related to biotechnology, 1982	184
7.1	Offshore supply contracts in British sector of North Sea, 1974–1980	222
7.2	British shares for each offshore supplies activity, 1974–1983	222
7.3	Summary of policies used in Britain, France, Norway and Holland	246

7.4	Effects of government policies on output and employment in the different sectors	251
7.5	Sectoral impacts of the policies used in the three countries	252
8.1	Top-ranking US electronics component producers 1955 and 1975	274

Figures

2.1 CAD applications strategies of selected suppliers 18

3.1 Diffusion patterns of CNC machine tools and industrial robots 55

5.1 Macrofilière and microfilière 137
5.2 The interreslationships (microfilières) within the various European and Canadaian videotex systems 139

6.1 Recombinant DNA: the technique of recombining genes from one species with those from another 165

Contributors

Margaret Sharp is a Senior Research Fellow at the Science Policy Research Unit. She is an economist specialising in industrial policy with experience both in universities and in government. She has recently been working on structural adjustment issues and inward investment policy in the United Kingdom.

Erik Arnold is a Research Fellow at the Science Policy Research Unit, working on policy, skill and employment aspects of technological change. Recent work includes studies on consumer electronics, computer aided design and government information technology policy.

Peter Senker worked in the electronics industry as an economist during the 1960s. During 1971 he was an IBM Fellow at Manchester Business School. He joined the Science Policy Research Unit in 1972 to lead a research programme on the skill implications of technological change and is now a Senior Research Fellow at the Unit.

Ernst-Jürgen Horn is an economist with the Institut für Weltwirtschaft in Kiel. He is a specialist on German adjustment problems and policies and trade relations with developing countries.

Henning Klodt is an economist with Institut für Weltwirtschaft in Kiel, which he joined in 1978. He is a specialist on issues of structural change, technology and productivity advance.

Christopher Saunders was a Professorial Fellow with the Sussex European Research Centre from 1973 to 1984 and is now a Visiting Professorial Fellow at the Science Policy Research Unit. He was previously Director of the National Institute of Economic and Social Research and Chief Economist with the ECE in Geneva.

Godefroy Dang Nguyen is a Research Fellow at the European University Institute, Florence. Trained as an engineer and economist, he worked with the Direction Générale des Télécommunications (DGT) in France before going to Florence to complete his doctorate on developments in the European telecommunications industry.

P. Lesley Cook is a Visiting Fellow at the Science Policy Research Unit and former Reader in Industrial Economics, University of Sussex. An expert on energy policy, Dr Cook has recently been working on the development of the offshore supplies industry in the United Kingdom.

Notes on Text

Reference to Germany in this text or tables means the Federal Republic of Germany throughout.

Table conventions

— Nil or negligible
n.a. Not available
( ) Estimate

Minor discrepancies in table totals and sub-totals are due to rounding.

Exchange rates

Conversions into dollars are made at current exchange rates for the year in question. Conversion rates are as follows:

$1 =	ECU	DM	F. Fr.	£	Lira	Neth. Guilder	S. Kron.	N. Kron.
1979	0.73	1.83	4.25	0.47	831	2.01	4.29	5.06
1980	0.71	1.82	4.23	0.43	856	1.99	4.23	4.93
1981	0.89	2.26	5.43	0.49	1,137	2.50	5.06	5.74
1982	1.03	2.43	6.57	0.57	1,353	2.67	6.23	6.45
1983	1.12	2.55	7.61	0.66	1,517	2.85	7.67	7.30
1984	1.26	2.85	8.94	0.74	1,757	3.20	8.27	8.16

Where a spread of years is used, the average for that period is calculated.

Sources: National Institute Economic Review; International Financial Statistics.

Preface

This book reports on a research project undertaken at the Sussex European Research Centre (now part of the Science Policy Research Unit) at the University of Sussex. The aims of the project were to study on a comparative basis across the main economies of Western Europe how new industrial activities were emerging and developing. It was conceived very much as a sequel to an earlier project which had studied the problems of adjustment faced by some of Europe's existing industries (see *Europe's Industries: Public and Private Strategies for Change*, edited by Geoffrey Shepherd, François Duchêne and Christopher Saunders, Frances Pinter, 1983). The idea with this second project was to highlight the new industrial activities which were emerging within Europe's economies and the comparative role played by firms and governments in this process.

As with the earlier study, these aims have been pursued through a series of case studies, each one focusing on some emergent new activity and attempting, by studying the interrelationships between firm and government strategy, to add a broader dimension to the narrow, self-sufficient sector study approach. More extended versions of these individual studies are being published in the Sussex European Papers series in the cases of computer aided design, biotechnology and telecommunications.

The project was sponsored by the Anglo-German Foundation for the Study of Industrial Society and the authors have benefited throughout from continuing support and advice from Barbara Beck and Hans Weiner of the Foundation, and specifically from a conference organised by the Foundation in Hanover in October 1984 at which the results of their research were presented and discussed with an audience of British and German business men, academics and civil servants.

The project has also benefited from (and added to) the network of collaborative relationships built up in earlier projects and now spanning economists and political scientists across Europe working on various aspects of industrial policy. Some of these have come together for the two series of workshop sessions organised for earlier stages of the project, others have provided advice, information and contacts for country visits and discussions. Besides the project authors appearing in this volume, particular thanks are due to Patrick Messerlin (Fondation Nationale des Sciences Politiques, Paris) Cor de Feyter

(Interuniversitaire Interfaculteit Bedrijfskunde, Delft), Rob van Tulder (University of Amsterdam), Paolo Cecchini (Commission of the European Communities), Giovanni Dosi (University of Venice), Joao Rendeiro (now with McKinsey, Madrid), and Geoffrey Shepherd, François Duchêne, Keith Pavitt, Luc Soete, Daniel Jones, John Surrey and William Walker amongst our colleagues at the University of Sussex.

The kind of research we have undertaken is only possible with a substantial amount of interviewing of business men and government officials. To all of those who have helped in this project the authors would like to extend their grateful thanks.

A word of thanks also for Ann Curry and Terrie Russell who, as successive secretaries to the project, have typed innumerable drafts with infinite patience.

Finally, I should like to add my own thanks to the project authors in this volume who have shown both patience and co-operation in the process of editing and finalising the text for publication.

Margaret Sharp
May 1985

1 Introduction

Margaret Sharp

This book is centred upon six case studies of new industrial activities in Europe. The purpose of each study is to throw light upon the process by which the activity has emerged and developed within the context of the European economies, and to point to relevant policy conclusions. Each provides a separate and in many senses unique picture of how the economic system adapts to new challenges and assimilates new opportunities. As a whole, they contribute to the general debate which is taking place about Europe's contribution within the world economy and in particular Europe's ability to compete with the technological prowess of the United States and Japan.

The terms of reference guiding the research project on which this book is based required a series of cross-country studies, focusing on the main industrial countries of Western Europe, which would 'contribute to an understanding of that part of the growth process which consists of new industrial activities'. A broad interpretation has been placed on these terms of reference. The general focus has been upon the way in which 'new industrial activities' have emerged and taken off within the context of the European industrial economies, but within this broad focus attention has been paid to a number of more specific questions. What provides the stimulus for development of the activity? What has been the respective role of government and industry, of large firms and small firms? Are there noticeable differences in performance between sectors or between countries? And if so, why? How does European performance compare with that of its competitors, particularly the United States and Japan? And what, if anything, can be done to improve that performance?

At this point it is perhaps useful to emphasise what the study is *not* about: it is not about the impact of new technologies upon employment. The project terms of reference did not require consideration of employment issues and even if they had done so, the question would have been difficult to handle within the case study framework, for employment is essentially a macro-economic issue. Micro-economic case studies can, if so directed, make predictions about the effect of new technologies on employment within particular sectors, but the main effect of new technologies is to increase productivity, which in turn increases real incomes which can then be spent on a very wide range of goods and services through the economy. Therefore, while productivity increases in one sector may, at least in the short term, reduce employment in that sector, they simultaneously increase demand (and employment) across a very broad range of industries. Thus, it is relatively easy, but totally misleading, to look at the overall employment issue through a series of micro-economic studies. Even had it been within the terms of reference of this project, there would have been little useful to say.

The concept of a new industrial activity

The case studies have been directed not towards explicit industries or sectors, nor in most cases towards specific technologies, but rather towards the looser concept of the 'new industrial activity'. This focus has both advantages and disadvantages. It has given a flexibility and breadth to the research which traditional sector or industry studies sometimes lack, encouraging a crossing of boundaries in pursuit of activities which are themselves reshaping and redefining these same boundaries. This captures part of the dynamic of the whole process of change. New industries do not emerge from the void; rather they emerge from existing industries, often from existing firms within those industries, as new ideas, new products and new processes are taken on board. The new activities of today are the new industries of tomorrow.

The drawback to focusing upon the concept of an activity is the paucity of statistical data. This problem arises precisely because of the degree to which the concept of an industrial activity crosses established industry boundaries (or at least industrial boundaries as defined by statisticians). The consequence is that much of the material used in these studies is derived not from the normal national and international statistical sources, but from consultant and trade association data quoted in newspaper reports and trade journals. The availability of such data varies from country to country and from sector to sector. But while pressure was put upon each contributor to quantify developments as far as possible, it was quite impossible in the circumstances to impose a standard form of presentation.

The choice of case studies

The choice of activities for the case studies reflected a number of factors. In the first place, the need for it to be 'new' meant that we were looking for areas where people were 'doing new things'— where new products or processes were being introduced and in the process either creating a brand-new activity (e.g. computer aided design, videotex) or transforming an older activity into what in effect was a new one (e.g. telecommunications). The pressures for change clearly came from two main sources. New technologies were opening up opportunities to produce new products or to transform old products and production processes; at the same time shifting patterns of taste and resource availability were creating 'demand side' pressures for change. It was important for the studies not to be wholly 'technology driven' and the decision to include the offshore supplies industry was explicitly so as to have within the project one study where pressure for change had come from demand factors—from the need after the first oil crisis of 1973-4 to develop North Sea oil and gas reserves as quickly as possible.

On the technology front, the impact of microelectronics was evident, but it was decided not to include microelectronics as a whole as a case study, partly because by 1982 (at the start of the project) microelectronics was already such a vast and unmanageable subject, and partly because a detailed study of the semiconductor industry had already been completed within the research group (Dosi (1981)). Rather, the decision was explicitly taken to move away from mainstream microelectronics to its applications, and four of the case studies reflect this decision. Two of them—computer aided design and videotex—are about new activities based on new products deriving from microelectronics, while telecommunications and advanced machine tools reflect the impact of microelectronics in transforming the process technologies on which these industries are based, in both cases the switch from electro-mechanical processes (which had originally been introduced some fifty years ago) to electronic controls.

With telecommunications, microelectronics is also transforming the product range. Instead of just one product, voice communication, it is making it possible for a plethora of new products such as videotex to be offered through the telephone system. Thus, the telecommunications industry, which for fifty years has seen little change other than the gradual but continuous expansion of the basic service, has been shaken on two fronts—technology and product range.

But there was another reason for choosing telecommunications as an area for study. The deregulation movement in the United States was by 1982 having its impact on the AT&T empire: until then the European public telephone authorities (PTTs) had seemed relatively invulnerable as public monopolies, although the new Conservative government in Britain already had plans to privatise British Telecom. It seemed likely that the mix of pressures and tensions arising from technical change, new products and deregulation, would provide fertile ground for investigation. The two chapters on telecommunications and videotex show that this is indeed the case.

Finally, it seemed important to reflect, in the group of studies, technological changes other than those stemming from microelectronics. Biotechnology was chosen, partly because, like microelectronics, it seemed to be an area of innovation which, in the long run, would have a very broad impact. But, unlike microelectronics, it was at a sufficiently early stage of development for the 'activity' to be manageable as a case study. It also had the added bonus of providing a case study where the early stages of commercialisation are currently in progress, it being almost possible to trace movement from research laboratory to scale production. The case study has the advantage of demonstrating not only the key role of firms within this process, but also the vital importance of academic science, and the need for there to be close linkage between the academic community and industry. Biotechnology is also an activity where venture capital finance has played a key role in 'take-off', highlighting differences in the capital markets of Europe and the United States.

The case studies: framework

As already explained, the focal point for each case study is the process of change: *how* the different activities considered have emerged and developed in the different European economies. Although each study develops its own story, the broad overall framework to which each adheres is as follows:

(i) what is the new industrial activity? How did it develop? What basic markets does it serve, etc? Are there any interesting interactions between emerging technologies and existing industrial structures?
(ii) what is happening in the main European countries? (In almost every case this includes the United Kingdom, France and West Germany, other Western European countries being included as appropriate.) Emphasis is put upon the strategies adopted

by the main 'actors'—firms, institutions (banks, universities, research institutes, etc.) and government—and the interplay between them;
(iii) the contrasting performance of the various European countries;
(iv) salient policy issues: the relative success of different strategies in the light of objectives; European as distinct from country-based policies; the case for co-ordinating policy at a European level;
(v) European performance in a world context—the relative competitiveness of Europe *vis-à-vis* the United States and Japan.

Although policy conclusions rank high amongst the issues to be covered, they should not be seen as the be-all and end-all of these studies. The unifying theme is that of 'the growth process' and it is the dynamics of this process which dominates, and which sets these studies apart from earlier studies on 'adjustment'. Too often such studies tend to adopt a comparative static framework of just two dimensions—before and after. Here the focus is upon the continuum of change. The main influence has come from those authors who look upon the economic system as evolutionary and upon the competitive process as an intrinsic part in that evolution: authors such as Penrose (1980), Downie (1958), Chandler (1977) and Williamson (1975) in relation to firms and markets, and more broadly from the innovation framework of Schumpeter (1939), and his modern exponents Freeman (1982) and Nelson and Winter (1982). This dynamic framework and its implications are discussed at some length in the conclusions (Chapter 8).

The case studies: a variety of interests and experience

While they begin from a broadly similar general framework, each case study develops its own *raison-d'être* reflecting both the interests and approach of the different authors and the contrasting experience in the different fields of activity.

Take, for example, Horn, Klodt and Saunders' study on machine tools and Dang Nguyen's on telecommunications. Both report on an older industry being shaken up by the advent of a new technology. The former shows how a fragmented industry was rudely awoken to developments by Japanese competition, survived by readjusting its product offering, and is now moving back into a position of strength. The moral they draw is that the process of adjustment is accelerated by allowing market forces to reign supreme and that government attempts at intervention have, in general, hindered rather than helped adjustment. By contrast, Dang Nguyen is looking at a highly concentrated, regulated and cosseted industry, dominated by the main public sector monopoly providers of the telephone network. Protection via public procurement has limited outside competition but the threat of entry into the equipment industry by the American majors, AT&T and IBM, has shaken complacency in the old order. Unlike Horn, Klodt and Saunders, Dang Nguyen does not depict an industry emerging from the crisis: the 'revolution' is only half over and it is not yet clear in what shape the industry will emerge. But he cautions against the view that free competition will be a panacea. Given the natural monopoly position of the network providers, he suggests, that the European route of the publicly operated digital network (the ISDN) has its logic and could, in the long run, offer a superior service.

Computer aided design (CAD) and videotex offer a similar contrast between private and public sector initiatives, but in this case concerning the establishment of a new product on the market. In one (CAD), American technology leads the field, while in videotex European technology is dominant, pioneered by the public sector firms, who nevertheless have difficulty in establishing a 'product recipe'. For CAD, the newly established American firms of the early 1970s quickly sifted out the 'product recipe', developing user-friendly software based on the microcomputer which enabled them to sell 'turnkey systems' for general application. Arnold and Senker argue that European firms are now at a disadvantage *vis-à-vis* their American counterparts because they failed to move in this direction quickly enough and in consequence now face considerable entry barriers established by the leading American firms, not least the tying of peripheral and software sales to specific hardware products. The growing sophistication of users and the need to readjust software packages for the new 32-bit minicomputers currently offers European 'little league' firms a chance to break into the 'big league' market—a chance which, they argue, will only be seized if action is co-ordinated at a European level.

The videotex example shows how difficult it is, even when the firms are large public enterprises, for such co-ordination to exist. Although Europe leads the way in videotex, the two major public firms involved—British Telecom and the French DGT—have long been vying with each other for domination of the world standard. Dang Nguyen argues strongly in the chapter on telecommunications for the initiative to be left to the enterprises themselves and removed from the political arena.

Biotechnology and the offshore supplies industries present a complete contrast to CAD and videotex. Far from being narrow, specialist areas of activity, both are very broad in coverage. Cook stresses the heterogeneity of the offshore supplies industry: the complex industry focusing on huge, one-off projects involving a massive networking and subcontracting operation, with the oil companies and major contractors holding the pivotal position within this network. Subcontracting has also been very important in biotechnology—at least for the American industry (much less so for the European) —and its motivation comes from similar causes, the need to spread risks (through fixed price contracts) and buy-in scarce expertise. But in general biotechnology to date has a much simpler, linear structure, based on the broad areas of development and the firms, upstream and downstream (in a technological sense) associated with these developments. In the United States the small firms play an important part in these groupings. In Europe it is the large chemical and pharmaceutical companies that are in pivotal positions and the key to the successful commercialisation of biotechnology in Europe lies very much with these companies.

Cook's main interest is in policy: what is the best means of promoting an activity such as the offshore supplies industry which is subject to fast but continuous incremental change? Whereas Sharp, in the chapter on biotechnology, urges the more extensive use of licensing and joint ventures, Cook warns against the British route of 'full and fair opportunity' with an open door to inward investment. She argues that while this has brought the major American contractors and subcontractors to set up subsidiaries in the United Kingdom, it has done little to encourage British subcontractors to become leading firms in the industry. By contrast, the French, albeit nationalistic, route of using major public firms (in this instance the oil companies) to foster a network of French suppliers upstream and downstream has had its successes. But this is an industry where risk plays a major part. To be successful, a firm needs to be financially strong and able to take a long-term strategic view of its own development. Nationalistic policies, such as those pursued by the French, may help to promote the necessary long-term confidence but they are in themselves inherently risky and often costly. Not all governments are prepared to accept this degree of risk.

The focus of the biotechnology chapter is on how scientific discovery is converted into commercial practice. Sharp contrasts the American scene with its very active small firm sector, with the almost total absence of small firm activity in Europe. Institutional considerations play some part in this: the venture capital market in Europe is new and still prefers funding American developments to European. But, particularly in Britain, it is growing in sophistication.

Sharp warns against the assumption that government activity can in itself create a new industrial activity, particularly in an area as fraught with uncertainty as biotechnology. Governments, she argues, can act as catalysts for new developments, but cannot and should not replace private initiative in the process of commercialisation.

The case studies: an overview

While each chapter stands on its own as a study in the development and growth of one particular activity, when all six are put together the question arises of the general conclusions to be drawn across the six. The concluding chapter describes the six preceding chapters as 'snapshots' of the process of change and points out that six snapshots do not make a movie. Rather than drawing conclusions, Chapter 8 identifies some common 'themes' which run through the case studies and tries to set them within a broader historical context.

What emerges is a picture of Europe at the crossroads of economic development. The future is a matter of choice. The common themes of Chapter 8—evolution not revolution; the pervasiveness of microelectronics; application and/or manufacture; the science-based industry; coping with uncertainty; and hares and tortoises—amplify this choice. Progress is evolutionary not revolutionary—most new industrial activities emerge not *de novo* but from existing firms and industries in response to demand and supply-side pressures upon them. It is the decisions (responses) of these firms that are crucial to the process of change. Equally, neither demand nor supply-side pressures are necessarily continuous; at present, for example, much of the stimulus for change is coming from the seismic changes taking place in microelectronics which, being a technology of very wide application, is sending 'shock waves' in many directions. But, in the long run, it is from application, not prowess in state-of-the-art techniques, that benefit derives. The question at issue is whether Europe has the capability and aptitude to respond to these supply-side pressures: to pick up and apply the new technologies as quickly and as readily as its competitors.

To answer this question at this stage in the book is to pre-empt the discussion in the final chapter. Let it suffice to say that although the case studies point to some shortcomings in Europe's capabilities (for example, the quality of the basic scientific infrastructure, the availability of skilled labour, the fragmentation of the R&D effort), most of these shortcomings are not insuperable. The real issue appears to be one of aptitude, or perhaps more particularly, attitude, rather than capability. It is the willingness on the part of firms and governments to respond to pressures and to take action to create capabilities that is questionable rather than Europe's ability, given these capabilities, to compete.

Bibliography

Chandler, A. (1977), *The Visible Hand*, Cambridge, Mass., Harvard University Press.
Dosi, G. (1981), *Technical Change and Survival: Europe's Semiconductor Industry*, Sussex European Papers No. 9, Sussex European Research Centre, University of Sussex.
Downie, J. (1958), *The Competitive Process*, London, Duckworth.
Freeman, C. (1982), *The Economics of Industrial Innovation*, 2nd edn, London, Frances Pinter.
Freeman, C. (1984), *Long Waves in the World Economy*, London, Frances Pinter.
Nelson, R. R. and Winter, S. G. (1982), *An Evolutionary Theory of Economic Growth*, Cambridge, Mass., the Belknap Press, Harvard University.
Penrose, E. (1980), *The Theory of the Growth of the Firm*, 2nd edn, Oxford, Basil Blackwell.
Schumpeter, J. A. (1939), *Business Cycles—A Theoretical, Historical and Statistical Analysis of the Capitalist Process*, 2 vols, New York, McGraw Hill.
Williamson, O. E. (1975), *Markets and Hierarchies: Analysis and Anti-Trust Implications*, New York, London, Macmillan.

2 Computer-aided design: Europe's role and American technology

Erik Arnold and Peter Senker

This chapter focuses on the type of computer aided design (CAD) known as 'interactive graphics CAD', which involves the production of engineering drawings through the use of a television-like screen on which pictures are drawn and displayed. The most significant advantage of interactive graphics is that it handles *pictures*. It involves interaction between the designer (with his screen) and the computer, which translates the design into mathematical data which it can store and retrieve. CAD can be applied wherever engineering drawings are made or used, and wherever there is a need for a visual analogue of design.

As a technology, CAD has been made possible by the enormous decline in the cost of computing since the early 1960s. It enables users to improve the quality of designs while simultaneously raising design and drawing productivity. It also brings important savings in the amount of time and labour needed to produce designs, provides scope for design improvements and increases the potential for standardisation in design and manufacture. The technology can be employed to draw or model objects in two or three dimensions, to modify drawings or models electronically, and to compile complete drawings from standard component drawings stored in a computer data base. It can also be used as a basis for numerically-controlled (NC) parts programming (writing the instructions to be followed by computer-controlled machine tools).

The potential importance of CAD in the overall computerisation of the engineering firm transcends the benefits it brings in its own right. Increasingly, in recent years, 'islands of automation' have been appearing in manufacturing firms: data processing; robots; computer-controlled machine tools; flexible manufacturing systems and other computer-based process controls. CAD is a significant 'island' of this sort, but its great advantage is that it can be connected to the others, for example to the firm's data processing systems to handle parts lists, bills of materials and to provide information for stock and accounting systems, and to computer-controlled machine tools and other production processes such as micro circuit and printed circuit board (PCB) manufacture.

As with other computer technologies, CAD is characterised by a very high rate of change and by the rapid growth of supplier firms. CAD technology is largely derived from the United States. Europeans have made important contributions, notably in advanced software, but have been far less successful than American-owned firms in packaging and selling their work. One result of this has been a lag in the adoption of CAD in Europe, compared with the United States, which adversely affects the competitiveness of European manufacturing firms *vis-à-vis* their American counterparts. Compared to the scale of effort in the United States, European governments have not been spending large sums of money. More particularly, individual European nations have each adopted their own policies, producing separate national reactions to the American lead and initiative in CAD technology. There has been little real European co-operation.

CAD technology and competition

ORIGINS

American government policy has been a crucial factor in the establishment and dominance of the American electronics and computer industries since the last World War. As with the start-up phases of other electronics technologies, military funding of CAD has been a key policy instrument. This led to a concentration of expertise in the major companies of the military-industrial complex, notably the aerospace companies. Procurement contracts as long ago as 1960 specified the use of numerically-controlled (NC) tools on given proportions of work done, and this was later to be the case with CAD, forcing the pace of adoption among these firms but also removing the risks of that adoption.

The first important use of interactive graphics technology was during the 1950s in the SAGE early warning radar system, where operators used light pens to point at targets shown on cathode ray tube (CRT) screens. Throughout the 1960s, space and military funding in the United States were important sources of new CAD techniques. But while military funding plays a continuing role even today, the pattern in CAD technology has followed that in microelectronics and the major thrust of change has moved from military to commercial markets.

By 1963, Ivan Sutherland was demonstrating that 2D interactive graphics CAD programs could be written and used in design and design analysis (Sutherland, 1963). By 1965, IBM, McDonnell and Boeing were all experimenting with graphics-based design using refresh screens driven by mainframe computers and linked to NC tools. This technology became known as 'CAD/CAM'—computer aided design/computer aided manufacture. Department of Defense (DoD) contracts, placed by the USAF with the major airframe builders during the 1960s, required the use of CAD for mechanical parts and led to the development of mainframe-based software. The most widely diffused was Lockheed's CADAM, the 1969 version of which has been licensed, developed and sold widely by IBM.

During the 1970s, the Department of Defense Manufacturing Technology programme continued to command considerable funding. In the period 1973–9, it was funded to the tune of $600m, and the budget for 1980 alone was $120m. Within the programme, the USAF's Integrated Computer Aided Manufacturing (ICAM) programme, concentrating on technology for batch production of high technology components, accounted for some $100m in the period 1978–83, and it has become the focus of the DoD's efforts to promote CAD applications to mechanical engineering.

Besides the defence industries, CAD in its early phase of development was used by two other industries. The American motor industry had an interest in CAD from the early 1960s. General Motors began work on a project called 'Design Augmented by Computers' in 1959, and was routinely using interactive graphics in design by 1963 (*The Engineer*, 19 February 1965). Ford was using interactive graphics by 1964, but was less advanced than GM (*Electronics*, 1 June 1964). By the late 1960s, Chrysler, Citroën, Renault, Pressed Steel Fisher and Fiat were also involved in CAD. Most of the interest was in digitising clay mock-ups of new body designs produced by the stylists and modellers because it helped make accurate dies. Spin-off to more general-purpose technology was minimal, but the motor industry is now a heavy user of bought-in CAD equipment. It is particularly proficient at modelling surfaces by computer.

By the late 1960s, interactive graphics also began to be used by the microelectronics industry as large and very large-scale integration (LSI and VLSI) made circuits so complex that it was impossible to keep track of designs without computer help. Today the American DoD's Very High Speed Integrated Circuit (VHSIC) program is providing American suppliers of CAD with a major source of competitive advantage in VLSI design.

The interactive graphics work of the 1960s was done on large 'mainframe' computers with refresh cathode ray tube (CRT) screens. A British visitor to MIT in 1968 remarked: 'large processors such as

IBM 360-67s appear to be almost two-a-penny' (Green, 1969). The need for these large computer facilities, combined with the heavy processing demands of refresh screens, made CAD at this stage of its development prohibitively expensive in most potential industrial applications. For example, a single-terminal system might cost as much as one million dollars.

THE 1970s

Around 1970, four crucial innovations appeared which formed the basis of the product recipe adopted by the successful CAD manufacturers of the 1970s. These were:

— cheap 'second generation' minicomputers
— the low-cost (Tektronix) storage tube
— structured programing
— virtual memory

Minicomputers were a crucial development because they drastically reduced the cost of computing power. The storage tube (CRT) was developed by Tektronix. It provided stable, flicker-free images and required far less computing power than the refresh tube previously used in CAD applications. Structured programing and virtual memory techniques were crucial because of the size and complexity of CAD software. Movement towards virtual memory techniques allowed large CAD programs to be stored and used in small computers with the minimum of penalty in computer response time. With the help of these techniques, minicomputers can be made to respond to CAD commands almost as fast as mainframes.

These newer, cheaper technologies permitted packaging of CAD know-how into turnkey systems, opening up the possibility of selling CAD hardware and software together as a complete package at a clearly defined all-in price. Up to this time, users had to assemble their own systems and write much, if not all, of the CAD software themselves. The turnkey formula provided CAD technology more or less 'off the shelf' to the vast potential market of 'naive' users, who would have been unable to construct their own systems.

A good deal of work on CAD in the American aerospace industry had been dropped in the 1969 recession, and this—together with the merger of McDonnell and Douglas in 1967, which resulted in the departure of much of the Douglas design teams—provided a pool of skilled manpower, some of whom were instrumental in establishing new CAD companies.

These new 'start-up' firms exploited the availability of people with CAD software skills by putting together 'turnkey' products: combinations of hardware and software which allowed their customers to buy CAD technology in a complete, packaged form. The normal pattern was to design a system for CAD of either mechanical or electronic components, and then to diversify into other applications areas as time went by. Thus, the activities of the turnkey vendors gradually brought together the three main streams of CAD technological development (aerospace, automobiles and microelectronics) of the 1960s. Soon, also, printed circuit board (PCB) design and cartographic applications were incorporated.

Most CAD suppliers can be classified into one of four categories, according to their origins. One set has military, aerospace or academic roots (these being somewhat intertwined in the United States, especially during the 1960s). Many firms in this group are at Cambridge, Massachusetts (the Boston suburb where MIT is located) and at Cambridge, England. In England, the role of the military is negligible.

A second group comprises firms in the electronics and scientific instrument sectors making computer-related equipment. Makers of CAD peripherals, such as Calcomp and Tektronix, have diversified into systems, partly in response to a perceived threat from Japanese entrants into CAD hardware. Some consultancies have also diversified into CAD systems supply, but their market shares are negligible.

A third group consists of firms or divisions of firms which sell heavily into military markets.

The fourth group is of established computer makers, such as IBM, CDC, Hitachi and Fujitsu, for whom CAD is a related new market. IBM is the only mainframe computer manufacturer to have significant market share at present. Recently, minicomputer manufacturers have begun to enter, with Prime Computer making the running. A substantial proportion of the computers used for CAD systems are made by Digital Electronics (DEC), but this company has not entered the systems market, although it launched graphics workstation hardware in 1983. The Japanese computer firms entering the market are usually from larger groups with widespread manufacturing interests, often including machine tools and robotics. In the long run a greater competitive threat to established American CAD suppliers may come from these Japanese companies than from American computer manufacturers such as IBM. Large R&D efforts are being made in robotics and machine tools as well as CAD, and products for CAD/CAM and Flexible Manufacturing Systems (FMS) are beginning to appear. The Japanese companies have substantial internal markets which they use as test-beds before selling to outsiders, so their products are likely to be thoroughly tested and to meet many users' needs. At present, however, Japanese competition in CAD is not significant.

During the 1970s, the turnkey manufacturers based their products on the 16-bit minicomputer technology of the beginning of that decade, notably DEC and Data General machines. By the end of the decade, however, minicomputer manufacturers began to make 32-bit 'super minicomputers', challenging the lower end of the mainframe manufacturers' product ranges. Like the 16-bit minicomputers before them, the 32-bit machines provided an important decrease in the cost of computing power, and were a natural choice of machine for new competitors entering the CAD supply industry from 1978/79 onwards. The prices of 16-bit computers had, of course, been falling through the decade, so that the intense efforts needed in the early 1970s to allow several graphics terminals to share a common processor were less necessary by the end of the decade because of the falling cost of that processor. Equally, the increasing use of CAD systems for design analysis, as well as for merely creating drawings, was stretching the computer power of 16-bit machines.

THE CURRENT SITUATION

At the beginning of the 1980s, both the shape of the hardware and the range of tasks involved in CAD were beginning to change. The pattern of the early 1970s, where a single 16-bit machine was shared between four or more terminals, began to give way to shared-logic systems using 32-bit processors and to more 'distributed' logic systems in which a number of workstations—each with its own 16- or 32-bit minicomputer—share access to a 32-bit supermini or a mainframe for processing-intensive tasks such as modelling and design analysis. The costs to the established turnkey manufacturers of coping with this change were considerable.

The world CAD industry is today dominated by a handful of American firms, of which the top seven held over 90 per cent of the market in 1980 (Table 2.1) and the market leader—Computervision —alone held 37 per cent. The industry has attracted many entrants, especially since the late 1970s, and by 1983 the seven leading firms had some 122 competitors, while still retaining their market share dominance. Growth rates of CAD suppliers have been high (Table 2.2), especially in the late 1970s, but have slowed down in the early 1980s, reflecting both the larger bases against which growth has to be measured and the effects of world recession.

In Europe, local suppliers compete with American manufacturers.

Table 2.1 World-wide turnkey CAD market shares

Firm	1979	1980	1981	1982	1983†
	(percentage)				
Computervision	33	37	30	27	25
Calma	15	12	11	12	12
Applicon	14	13	9	8	6
Intergraph	9	11	10	13	15
Auto-trol	11	10	5	4	3
Gerber	4	3	*	*	*
IBM	5	6	16	19	21
Other	9	8	19	17	18
Total market ($m)	310	510	894	1,207	1,600

* Included in 'other'.
† Estimated.
Source: Kurlak, 1982, 1983; press reports.

The penetration of the American market by foreign firms is negligible, whereas American suppliers tend to dominate the markets in which they sell. CAD suppliers also compete with in-house systems developed by users, such as Siemens. Kaplinsky estimated the global value of CAD 'sales', including such in-house systems at $1bn for 1980, and cites an estimate of 4,265 systems (excluding IBM installations) for the world-wide installed base of CAD systems by the end of the 1980s. These systems were concentrated in the industrialised countries (Kaplinsky, 1982).

Following the typology developed by Sciberras (1977) to describe semiconductor firms' strategies, CAD firms' product strategies may be described as 'big league' or 'little league'. The dominant American turnkey suppliers follow big league strategies. Each big league firm operates across several of the specialist applications areas into which the market is divided. These specialisms include PCB design, LSI design, 2D and 3D mechanical design, architecture and civil engineering, cartography. Little league firms are smaller than big leaguers. They choose one or two specialist CAD applications and operate solely within them, thus limiting the extent of the market which their products address. Figure 2.1 illustrates how, in practice, CAD suppliers operate along these two dimensions: the proportion of the CAD systems supplied; and the generality of application. The majority of firms cluster at the left of the diagram, but their combined market share is small. The dominant firms cluster on the right.

Table 2.2 Global sales of CAD/CAM turnkey systems by major United States vendors ($000s) 1969–1983

	Date of origin	1969	1971	1973	1975	1977	1979	1981	1982	1983
Applicon	1968					16,640	28,469	96,000	96,000	100,000
Auto-trol	1959		±1		4,835	12,549	33,540	48,000	44,000	49,000
Calma	1964				9,000	17,000	43,000	100,000	140,000	195,000
Computervision	1969	51	2,567	8,510	14,572	28,188	103,004	271,000	325,000	395,000
Gerber	1974					1,400	10,200	18,000	19,000	
IBM	1974					20,000	40,000	145,000	225,000	340,000
Intergraph	1969	50	480	920	3,200	9,173	29,518	91,000	156,000	246,000
Unigraphics	1961						7,500	35,000	46,000	60,000
Other							21,000	68,000	156,000	215,000
Total						110,000	316,231	872,000	1,207,000	1,600,000

Sources: Kaplinsky, 1982; Kurlak, 1982, 1983.

18 Erik Arnold and Peter Senker

	Numerically controlled	Mechanical	Printed circuit board	Integrated circuits	Piping	Architecture and civil engineering
US Big League						
Computervision	*	*	*	*	*	*
Applicon	*	*	*	*	*	*
Intergraph	*	*	*	*	*	*
Autotrol	*	*	*	*	*	
Calma	*	*	*	*		
Gerber	*	*	*			
Unigraphics	*	*				
IBM	*	*				
European Little League						
Ferranti	*	*				
CIS	*	*				
Matra		*				
Aristo	*	*				
Kongsberg	*	*				
Racal			*			
Quest			*			
Secmai			*			
CAL			*			
ICAN			*			
Compeda					*	
GMW						*
ARC						*

Figure 2.1 CAD applications strategies of selected suppliers

The firms in the middle tend to be computer manufacturers who have diversified into CAD, with IBM and Prime being the most significant.

To date, there is little evidence of European CAD firms wishing to grow into multinationals. In the United Kingdom, Racal (and Quest when it existed) exports the greater part of its (PCB-orientated) production, in contrast to American manufacturers for whom exports provide less than half of sales, while Cambridge Interactive Systems (CIS) and Ferranti both became, by British standards, substantial exporters of mechanical design soft- and hardware in the early 1980s (CIS have since become a part of Computervision). Norwegian suppliers have been forced by the small size of the domestic market to establish sales facilities abroad (and, in the case of Kongsberg, R&D facilities also). The French Matra Datavision company has set up marketing facilities abroad, but other French software is marketed by American multinational companies.

Government policies in individual countries

FRANCE

In French, a distinction is made between *Conception à l'Aide d'Ordinateur* (CAO) and *Dessin à l'Aide d'Ordinateur* (DAO). Essentially, this corresponds to the distinction between modelling and computer aided drawing. French-written software is almost wholly concerned with CAO and French interest in CAO has been associated with the problems of modelling complex surfaces which cannot be generated in mathematically simple ways. This type of surface modelling is required in designs which have an aero- or hydrodynamic aspect, such as airframes, car bodies, and ships' hulls. The benefits of CAO are in better conceptualisation and optimisation of designs rather than in productivity gains, which are the province of DAO.

CAD was originally introduced in France between 1964 and 1970 in four sectors: aeronautical engineering; electronics; cars and shipbuilding. Poitou (1983) argues that the main thrust for the use of CAD came from the use of NC machine tools, the growing complexity and variety of products, quality requirements and reduction of design delays. Although CAD led to reduced design costs, this was not the main factor determining its adoption.

NC machine tools required rigour and precision in the preparation of tooling and in the production process itself. Companies with the resources to apply computers to do this did so. Thus Citroën undertook what it called the 'mathematisation of the product' from 1965. The first studies at Renault which were to culminate in the UNISURF CAD software began in 1964 with research on the representation of sculptured surfaces as a basis for NC program preparation. Bezier's work associated with this project produced what has become the dominant technique for the representation of complex surfaces: so-called 'Bezier's patches'.

The aerospace industry was prompted to develop CAD systems largely by the need to cut lead times. In shipbuilding, the Chantiers de la Seyne yard sought in the 1960s to be able to produce ships more or less 'on demand' by exploiting a CAD data base, in response to American and Japanese domination of series-produced ships. The motivation leading firms to develop CAD systems was to overcome technical bottlenecks and problems, rather than to increase productivity. CAD developments in the aerospace industry were seen explicitly as being enforced by the use of the technology by American competitors. In all cases, the number of graphics terminals used was low. Some tracers were displaced, but this was a side-effect of the use of CAD technology, rather than an objective.

Currently, the aerospace industry uses French software for high-specification 3D systems, and buys American 2D and 3D systems as well. In general, French systems are preferred at the conception and design stage (CAO) and for surface modelling, while American turnkey systems are used in designing and drawing components. The pattern is similar in the automobile industry. Poitou reports a tendency for the smallest CAD users (less than 100 employees) to develop their own systems. He does not say whether these are graphics-based (Poitou, 1983).

France accounted for about 6 per cent of world CAD installations in 1980, or 17 per cent of European installations. The French market is dominated by American suppliers, notably Computervision, Applicon and IBM who account for something like 80 per cent of sales (interviews). Our estimates of the stock of CAD systems in France at about the end of 1982—based on others' partial accounts—are given in Table 2.3. In the PCB design area, the French Secmai and the British Racal companies are important.

French CAD suppliers tend to supply CAO, rather than DAO, software and to follow little league, niche strategies, since they are marketing products which are spin-offs from their other activities in aerospace and automobiles. The principal academic contribution to CAD in France has been EUCLID, which was developed for complex 3D modelling by the Centre Nationale de Recherche Scientifique (CNRS). The indigenous French presence now includes

Table 2.3 French CAD market, estimates of shares of installed stock, 1982 (excluding IBM)

Supplier	Number	Share (%)
Applicon (US)	93	21
Assigraph (Fr)	4	1
Autotrol (US)	15	3
Calma (US)	10	2
Cisi (Fr)	12	3
Computervision (US)	160	36
Gixi (Fr)	6	1
Graphael (Fr)	13	3
Matra Datavision* (Fr)	25	6
Racal Redact† (UK)	50	11
Secmai (Fr)	53	12
Storm (Fr)	4	1
Other	5	1
Total	450	100

* Interview.
† *Lettre 2000*
Primary source: Poitou, 1983. No data available for IBM.

the Datavision subsidiary of Matra, which was set up in 1979 to market and continue the development of the EUCLID software. Datavision attaches great importance to its claim that EUCLID is portable between computers, and it has been implemented on eight different manufacturers' hardware. EUCLID is widely used in the international aerospace industry and Datavision claims half the French market for 3D software.

Dassault's CATIA—another 3D system for aerospace application—is marketed by IBM, and has been acquired by several American, European and Japanese companies. Adoption of CATIA by IBM is widely cited as demonstrating that France is surpassed only by the United States in CAD software capability. Aerospatiale's SIGMA and SYSTRID are marketed respectively by Battelle and CDC.

French government intervention in computing began after 1965, when the American-controlled company Machines Bull refused to supply computers to the French nuclear weapons programme. The 'Plan Calcul' was set up in 1966 with the intention of establishing a purely French computer industry. Under the aegis of the Plan Calcul, the 'Plan Composants' was established in the same year to co-ordinate French efforts on semiconductor component technology. In 1975, the 'Plan Mini-Informatique' (minicomputers) and the 'Plan Peri-Informatique' (computer peripherals) were also established within the overall Plan Calcul. However, these successive *Plans* have not seen great success. The inability of the French computer industry to maintain a presence in mainframes led to the invitation to Honeywell, in 1976, to take a 47 per cent stake in a new joint venture: CII-Honeywell Bull. This created a very large national firm, which accounted for some 12.5 per cent of the European mainframe market in 1978 (DEK, 1980). The reorganisation of the computer industry cost a little under 3bn frs. (approximately $670m) in the period 1975-9.

In 1979 the 'Informatique et Société, Premier Plan' was launched in response to the Nora and Minc report on electronics and society (1978). In 1982, the French government decided to launch a major programme to promote the '*filière électronique*'. While CAD was not identified specifically in the initial policy document, the strong position of the French industry in CAD software was considered important.

The CNRS, the French science research funding body, has played a role in funding CAD research in the university sector and elsewhere. On the user side, the Association Française Mission pour la Conception et le Dessin Assisté par Ordinateur (MICADO) was set up in 1974 as a non-profit making body to promote CAD adoption, and has received co-operation and financial support from the Ministry of Industry. MICADO comprises 140 representative members of public bodies, universities and research centres, manufacturers and marketers of information technologies, and industrial users and potential users of CAD. Conferences and seminars are organised, and MICADO catalogues and demonstrates CAD hardware and software at its headquarters at Meylan, near Grenoble. As well as providing bulletins to its members, MICADO also promotes sector-based activities in conjunction with the Agencie de l'Informatique and diffuses software emerging from university research.

Nevertheless, the concentration of French products in CAO, as opposed to DAO, means that the size of the market available to French companies is very limited. The ratio of CAO to DAO terminals in France is—as elsewhere—very low, so that the American products which have been associated with drawing office automation (DAO) dominate the CAD market overall. Note the relatively small share of the market taken by French firms in Table 2.3.

The lack of a French 32-bit minicomputer makes combined hardware-software strategies difficult to follow. A good deal of weight, therefore, is laid by suppliers of French sophisticated CAD software on software portability. This, however, makes them vulnerable to the whims of computer manufacturers, particularly IBM. As a result, despite the high quality of French work in state-of-the-art techniques, this software-based strategy combined with stress on CAO at the expense of DAO guarantees that French industry is restricted to the little league of CAD.

WEST GERMANY

CAD came into use in the West German automobile, aerospace and electronics industries from the beginning of the 1970s. Brown Boveri, SEL, Masserschmidt and Bosch were among the earlier users.

In 1981, the number of installations of all types of CAD/CAM systems was about 300, of which 80 per cent involved interactive graphics. The trend has been for turnkey and graphics-based systems to grow at the expense of the shares of other types. (See Table 2.4.) The use of these CAD/CAM systems in 1981 was approximately 40 per cent for mechanical applications, 40 per cent for electronics, and 20 per cent for construction and cartographic work (interview dicussions).

Sales of *turnkey* CAD systems did not begin to be made in substantial numbers until 1978/9, and the stock of CAD systems was about 100 in 1980 (Roth, 1982). Research commissioned by Calma suggested that this had risen to 200 systems by the end of 1982 (Lewandowski, 1983).

The automobile industry is both an important user of CAD technology and a major area for potential growth. West Germany appears to have some of the largest users (i.e. large firms with multiple units) of CAD outside the United States. The largest is in the aerospace industry, having 100 terminals and 600 trained operators.

The German market has recently been turning away from American turnkey-type products, which are seen as coupled to the drawing function, and moving towards more integrated or integrable systems —notably CIS's MEDUSA. People are now more interested in connecting CAD into a broader system of automation than used to be the case. Equally, they want things they can add to their existing hardware, and the major American firms' products have often been very difficult to integrate with other systems. German (DIN) standards are said to be hard to implement on turnkey equipment, especially in mechanical engineering, and this makes the US suppliers vulnerable.

Table 2.4 Estimates of West German CAD park, 1981, shares of supplies (excluding IBM and Calma)

	Number	Share (%)
Applicon (US)	20	10
Aristo (US)	5	2
Autotrol (US)	10	5
Calcomp (US)	1	*
CDC (US)	15	7
Computervision (US)	67	32
Dietz (FRG)	3	1
Gerber (US)	38	18
Kongsberg (N)	3	1
McAuto Unigraphics (US)	1	*
Intergraph (US)	3	1
Prime/CIS (US/UK)	16	8
Siemens CADIS (FRG)	17+	8
Summagraf (US)	10	5
Total	209	

* Negligible.
† 70 workstations; number of systems not given. Four workstations per system have been assumed.
Source: Industry discussions.

The relative sophistication of much West German use of CAD may be explained by the greater availability of highly trained engineering personnel than in, say, the United Kingdom (Prais, 1981). Between a third and a half of the CAD/CAM user firms interviewed by the Department of Applied Systems Analysis (AFAS) at Karlsruhe (see Reihm and Wingert, 1983; Wingert *et al.*, 1982) have employed at least one person with CAD system development experience from college. As a result, a great deal of system development knowledge can be incorporated in CAD implementation.

The German CAD market is dominated by foreign suppliers, mostly from the United States, but with British firms also playing an important part. IBM has recently become a major force—unfortunately too recently for it to show in the sales figures quoted in Table 2.4—with many sales to their existing mainframe users. Computervision is strong in electronics and electrotechnics. American companies do less well in the German mechanical design market where implementing German drawing standards and language in their systems is more important. From 1983, demand in mechanical applications for MEDUSA from CIS (recently taken over from its British founders by Computervision) had tended to exceed that for the products of Computervision itself. Particular efforts appear to have been put into adapting MEDUSA to the German market, with a trial installation in a German partner firm to get the German standards right. The British Ferranti company had installed about 60 of its CAM-X workstations in West Germany by 1983. Thus, the recently successful suppliers of mechanical design equipment in Germany were firms which based their product on technical data bases and solid modellers.

West German firms supplying CAD systems include Nestler and Kuhlmann, which are traditional drawing equipment suppliers. Their approach is to produce equipment which can be developed naturally from existing drawing products, using drawing-board sized digitizers and 16-bit computers. This approach pre-dates the CAD developments of the 1960s, and it is difficult to imagine it succeeding now.

Of the old-established drawing equipment firms involved in CAD, Aristo is probably the most important. In 1979, Aristo bought a New York firm called Information Displays Inc. (IDI), which produced a 2D mechanical design system. Aristo expanded the IDI mechanical system to take the user right through to NC, and implemented it in German. The product was limited, partly because it was based on 16-bit technology and this held back sales until it was replaced with a 32-bit system.

Siemens is an important manufacturer of CAD in Germany. Its product is called CADIS and comes in 2D and 3D versions but runs only on Siemens' computers. However, to date, all 31 users worldwide are within Siemens, 24 of them in Germany.

Dietz is another established German computer manufacturer which is diversifying into CAD via its 'Technovision' subsidiary. It began CAD development in 1977. There is an architectural system with an alphanumeric interface, as well as 2D and 3D interactive graphics products. The software design team is small—5 engineers in 1980, with 15 software man-years embodied in the product to that point.

German government policy for CAD needs to be seen against a background of a relatively strong domestic computer industry, led by Siemens, competing in a market dominated by IBM. The major source of funding for CAD development is the Bundesministerium für Forschung und Technologie (BMFT)—the Ministry of Research and Technology. Its activities in electronics-related sectors have taken place under four programmes:

- data processing (*Datenverarbeitung*);
- electronic components (*Elektronik*);
- key technologies (*Technologische Schlüsselbereiche*); and
- humanisation of work (*Humanisierung des Arbeitslebens*).

Of these, the data processing programme has had the most influence on CAD. Most programmes or sub-programmes are administered on behalf of the BMFT by 'Projektträger', which evaluate proposals and monitor the progress of research. The Projektträgerschaft for mechanical engineering CAD/CAM is at the Atomic Research Centre in Karlsruhe and serves as the channel both for funding CAD developments in the Hochschülen and for the Federal Government's measures to promote adoption.

The data processing programme was in three phases. The first two, lasting from 1967 to 1975, were principally concerned with developing computer technology, especially hardware. The third phase, 1976 to 1979, concentrated more on minicomputers, process control computing and computer applications, including CAD. Projects supported by this BMFT programme include: work on geometric modelling by IKO Software Service ($1.34m—DM 29.m); an integrated programme for shipbuilding by Forschungszentrum des Deutschen Schiffbaus ($1.61m—DM 3.5m) and development of a building system by Rechnen und Entwicklungsinstiüt für EDV in Bauwesen ($1.52m—DM 3.3m). Individual Fraunhöfer Institutes have also played a role in CAD applications.

In policy terms, CAD has not been seen as a discipline distinct from data processing until very recently. In 1979, at the end of the Datenverarbeitung programmes discussed above, it was decided that the development and use of CAD/CAM software needed promoting. Previously, efforts had been decentralised and were specific to industrial sectors. From 1979, they were put together with a budget of DM 50m ($27m) per year. Half of this budget was to be found from industry, and half from the state. CAD software development work done at the Hochschülen was to be funded 75 per cent by the state and 25 per cent by industry, in order to ensure that only things which had an immediate industrial user were developed.

The Christian Democrat Government, which came into power in 1982, believes that CAD/CAM support schemes have been too *directive* of industry, and a new support programme has been agreed by the Cabinet. This involves allocating DM 530m between 1984 and 1988 to provide financial support to firms wanting to start using CAD, CAM, CAPlanning, or robots.

A CAD demonstration laboratory has been set up in the centre of

Karlsruhe, and the Projektträgerschaft staff are also involved in presenting CAD at seminars and congresses to increase awareness in the Federal Republic. The Verein Deutscher Ingenieure (Association of German Engineers) in Berlin has about 70 people administering a separate scheme for electronics and microelectronics, but this scheme is more about new CAD product development than fostering awareness and use of the technology.

Despite a level of support of CAD technology consistent with generating and maintaining national capability in CAD technology, there is little substantial German activity on the CAD supply side. None the less, the use of CAD appears sophisticated and may be better integrated with other aspects of computer aided engineering and production control than in some other countries. The importance of the national automotive industry, including the subsidiaries of American multinationals, in the German economy, may be one factor explaining this, as high-volume car makers must necessarily be relatively sophisticated in their use of computers in order to stay in business. This is perhaps particularly true for manufacturers of the more expensive types of car favoured in West Germany. The high market share of mainframe computers held by IBM in Germany allows German users ready access to IBM's particularly integrated approach to manufacturing industry computing and this may in itself be a factor promoting a particular desire for interconnectedness between CAD and other related technologies. Another factor involved in national capability in using CAD and related technologies is the level and type of training of skilled and professional manpower.

THE UNITED KINGDOM

The United Kingdom had the beginnings of a strong CAD supply industry in the 1960s. As in the United States, the main customer in the early stages of development was the aerospace industry, but a number of other industries, including civil engineering and automobiles, were showing interest in developing systems by the end of the decade and there were four firms—Elliott Automation, Ferranti, ICL and Marconi—all offering CAD systems or services based on their own interactive graphics hardware, and Racal was offering a PCB system based on Elliott Automation hardware. Meanwhile, in 1967, the Ministry of Technology had set up the Computer Aided Design Centre (CADC) in Cambridge, aimed at creating national capability in the technology.

The reorganisation of the British computer industry in the late

1960s and the formation of ICL (which included many of the firms involved with CAD) killed much of this early initiative. For one reason or another, Elliott Automation, Marconi and Ferranti all withdrew from the market (Elliott's withdrawal also affected Racal) just at the point when technological innovations were making the marketing of 'turnkey' systems feasible. The result was that the market remained limited and highly specialised (with users tending to develop their own systems) until 1977, when the American firms began to penetrate the British market.

The installation in 1977 of a system at Baker Perkins in Peterborough, which was supported by a substantial grant from the Department of Industry, marked the beginning of this growth phase. By mid-1981, it was estimated that there were some 150 to 200 CAD systems installed in the British engineering industry, and that the number of substantive systems in operation was approximately 400. Market shares for 1981 are shown in Table 2.5, which also demonstrates the current dominance of American turnkey suppliers. The market has been growing fast. At mid-1983, the Department of Trade and Industry (DTI) estimated that the total CAD park stood at 1,000 systems (but their estimates include systems based on microcomputers, which were excluded from Table 2.5). One of the success stories of recent years has been Cambridge Interactive Systems (CIS) which, in 1981/82, became an important competitor in CAD for mechanical design. However, Computervision bought CIS in 1982, capturing this important challenge to its market dominance in the United Kingdom and other European countries.

The strengths of the British CAD industry are in PCB design, where domestic manufacturers have a substantial share (probably the greater part) of British installations, in process plant design and in solid modelling. The British suppliers of PCB design equipment are Racal and Quest. During the 1970s, the Ministry of Defence funded the implementation of Racal's software on PDP-11 minicomputers, and the resulting turnkey product has formed the basis of Racal's successful growth, with the Ministry receiving a royalty on sales. More than 70 per cent of production was being exported in the early 1980s. Quest was established in the late 1960s by ex-Plessey personnel. The first product was a plotter for PCB manufacture, followed by digitizers and PCB design systems. Financial problems during the last recession (it had tried to diversify into architectural applications by buying Genesys in 1981 and had launched its own 16-bit minicomputer hard on the heels of Computervision's 32-bit machine) forced it into the hands of the receiver in 1983 before being sold to its largest British customer—GEC Marconi.

Table 2.5 British turnkey CAD park, market shares, 1981

Supplier	CPU-equivalents	Share %
Applicon	14	3
Autotrol	14	3
Calma	52	13
Compeda	22	5
CIS	10	2
Computervision	120	29
Ferranti	9	2
Gerber	2	
IBM	50	12
Intergraph	6	1
Lundy	6	1
Quest	40	10
Racal	60	15
SDL (AD2000)	1	
Tektronix	1	
Unigraphics	7	2
Total	414	100

Notes: Gerber excludes clothing applications. IBM had about 200 terminals, equivalent in terms of minicomputer power to about 50 CPUs. Quest and Racal supply small systems, so these figures overestimate their importance. The Tektronix figure is an underestimate. Architectural, cartographic and shipbuilding systems are excluded, as are microcomputer-based systems.
Source: Industry discussions.

Other British firms worth mentioning are:

(i) Shape Data, which was established in 1974 by four people leaving the CAD group of the Cambridge University Computer Laboratory. The firm's ROMULUS was one of the first solid modellers to reach commercial applicability. It was taken over by the American company Evans and Sutherland in 1982.
(ii) Ferranti, which re-entered the CAD industry in 1980 with CAD/CAM software and workstations for mechanical design, based on the ROMULUS solid modeller.
(iii) Compeda, which was a venture established by the National Research Development Corporation (NRDC) in 1977 to market British-produced software. The firm had no control over the type of products it would be offered (largely by academics) and it failed to develop a coherent product range.

Losses in 1982 led to a takeover by the American Prime Computer Company, after a bid by Computervision had been blocked by the DoI. The PDMS process plant design system, designed by CADC and Isopipe, and marketed until 1982 by Compeda, is firmly established as pre-eminent in its field worldwide, but its market is limited by the special problems it addresses and by its unfriendly user interface.

No British firm offers equipment to compete with the breadth of capability of American turnkey systems. Indeed, until very recently, British firms outside the PCB applications sector have not sold packaged systems at all, and have therefore largely been excluded from the crucial first-time buyer market. Effectively, therefore, manufacturers operate in the little league of CAD supply. Certain firms have had considerable success because of the technical strengths of their little league products. This strength has sometimes (e.g. CIS) related to better connection from CAD to CAM than can be provided by the dominant turnkey suppliers within the relevant niche. In general, British firms have remained highly specialised.

Our study of CAD users in the United Kingdom (Arnold and Senker, 1982) found little evidence outside the electronics industries of use of CAD being integrated with other forms of computerisation within firms. By and large, American-made turnkey systems were being used within the drawing office, where they had been cost-justified in terms of labour savings, to increase the productivity of draughtsmen and related clerical staffs. Only in one company (in the aerospace industry) was the need to connect CAD with other computer technologies leading to organisational change.

While British policymakers were very early in adopting positions on CAD, these have sometimes been contradictory and responsibility for policy making which affects CAD has been diffused. As we have seen, the Ministry of Technology was responsible for the establishment of the CADC at Cambridge (1969) which was equipped with an Atlas 2 computer—for its time, a formidable machine. Initially, the Centre was funded wholly by the Ministry of Technology (later absorbed into the Department of Industry—DoI), but by 1982 about half of the £4 million funding to support the Centre's 150 or so staff was coming through software sales and consultancy (*The Financial Times*, 2 April 1983) and it was sold in 1983 to a consortium comprising ICL, the computer bureau SIA, W. S. Atkins and Cambridge University. Throughout its existence, the CADC had been a source of personnel for start-up firms—notably CIS and Shape Data. But despite the availability of some CAD expertise in the United Kingdom, there were no significant British supplier start-ups during the early to mid-1970s when American turnkey suppliers were establishing and dominating the CAD industry, and no significant diffusion of CAD into British industry. In this sense, the response to these initiatives has been disappointing.

Another area of policy that very much affects CAD is policy towards the computer industry. We noted above that many of the British computer companies involved in CAD at the end of the 1960s were merged to form ICL (in which the government took a 20 per cent stake) in 1968. Those that were not merged were barred from the markets for large computers by the terms of that merger. While the ICL merger and subsequent government support have enabled a continuing British presence in the mainframe computer market, overall it probably damaged British prospects in CAD because it focused the activity of the merger companies on mainframe computers rather than on other applications, while encouraging those not involved in the merger to move out of the computer field altogether.

By the end of the 1970s, the government was becoming increasingly concerned at Britain's relatively poor performance in the microelectronics field and, particularly, at the lack of awareness in industry about potential applications of microelectronics. From 1978, government support for microelectronics technology generation and adoption increased considerably, and a number of special support schemes for microelectronics were introduced. Some of these schemes, in particular the provision of consultancy through the Microprocessor Application Project (MAP), could be used in connection with CAD adoption. In January 1982, the DoI announced a special 'CAD/CAM' scheme with £6m ($10.5m) of funding to promote the adoption of CAD in mechanical engineering. A similar scheme to cover computer aided design, manufacture and testing (CADMAT) in the electronics industry was announced later in the same year, and funded with £9m ($15.7m). A further £12m ($21m) was allocated to these schemes in July 1982 and another tranche of £10m was added in 1983 with the remit braoden to include computer aided production management. In addition, the DoI has supported six 'hands-on' CAD experience centres, where potential users can try the technology. Although these schemes have helped to diffuse knowledge of CAD and its applications, they have done little to help the British industry itself. Early schemes (particularly the MAP consultancies), with their emphasis on application, tended to favour the 'user-friendly' American turnkey systems. Recent measures to subsidise CAD adoption (the DoI's CAD/CAM and CADMAT schemes) have favoured British equipment, but this has led to the acquisition of often specialised or low specification equipment because of the limited range of British equipment available.

On the supply side, the DoI attempted to foster joint activity or merger between Racal, Compeda and Quest in 1981/82. After the failure of this initiative, a substantial proportion of the British pool of CAD-skilled people was allowed to fall into foreign hands, making the subsequent appearance of an important role for British firms unlikely. CIS was taken over by Computervision; Shape Data was taken over by Evans and Sutherland; and a substantial interest in Quest was sold abroad before that company went bankrupt. The sale of the CADC to a private consortium of British interests in 1983 effectively marked the abandonment of state influence on the supply side of the British CAD industry.

SWEDEN

CAD systems started to appear in Sweden in the early to mid-1970s. Most early users were large firms, predominantly in electronics (Computers and Electronics Commission, 1981). At the beginning of the 1970s, Kockums Shipyards, SAAB-Scania and Volvo had CAD systems which they had developed in-house. The robot maker ASEA has since developed CAD systems for both mechanical and electronic design in-house, and Jacobsson and Widmark have developed their own architecture/civil engineering system.

In 1979, Sweden had some 60 CAD installations, split between the 22 largest NC machine tool/robot users who had 40 installations, and 20 in the rest of industry (DEK, 1981). In 1979, L. M. Ericsson, ASEA, Volvo, Sandvik, SAAB-Scania and Electrolux together had 25 per cent of the Swedish NC/machine tool stock, 40 per cent of industrial robots and 50 per cent of CAD systems (Carlsson, 1982). Clearly, these larger enterprises play an important role as pioneer users of new manufacturing technologies. The average number of people working with one CAD system is now twelve, according to the trade union SIF. SIF has classified Swedish CAD installations in 1982 by size of firm and type of application, and this classification is reproduced as Table 2.6.

In general, the Swedish pattern is now one of very large users on the one hand (some thirty in mechanical engineering and twenty in electronics), and a number of very small ones concentrated in PCB, bureau and architectural-type activity on the other. The large firms are moving into a period of expanding their CAD capacity after a lengthy pilot phase. Firms do not tend to co-ordinate their CAD

Table 2.6 CAD/CAM—number and type of users in Sweden, 1982

	Total	M	E	M + E	ACE	SB
Large firms	54	24	11	8	11	0
Small and medium	47	1	10	0	14	22
Industry total	101	25	21	8	25	22
Technical schools, etc.	(6)*	n.a.	n.a.	n.a.	6	n.a.
Other (e.g. local government)	7					
Total known users	114					

M = mechanical applications; E = electronics applications; M + E = both mechanical and electronics applications; ACE = architecture, civil engineering and construction; SB = service bureaux.
* ACE use. Other figures not available.
Source: Lindberg (1983).

acquisition decisions, but take these at plant level, with top management not being particularly involved. Service bureau activity is growing rapidly, with small firms being attracted to this method of using CAD since it avoids the capital cost of systems acquisition. In architecture and civil engineering, many users are small and often rent equipment or share it with other comparable firms (Lindberg, 1983).

Users in Sweden are becoming more sophisticated in their approach to the technology. The time taken by individual firms between a decision to buy CAD and an order being placed has tended to rise to about two years (compared with about a year in the United Kingdom), reflecting potential users' increasingly demanding attempts to measure and compare the characteristics of different suppliers' systems.

The dominant American turnkey manufacturers began to make significant numbers of sales in the period 1978-80. Many of these were pilot installations, and CAD sales in 1981 were low. CAD suppliers report some disappointment with the productivity gains from these pilot systems bought in 1978/9 and this probably combined with recession to deaden the CAD market. Design in Sweden is not subject to much division of labour, the 'konstrukteur' tending to take designs from the analysis stage to the finished drawing, with less tendency to split drawing off from other design functions. This could reduce the productivity gains from CAD.

By the summer of 1982, there were about 205 systems installed in Sweden and it is seen as a particularly promising market. Some

25 different companies have sold CAD systems in Sweden, and another 10 to 15 are trying to market them there (Lindberg, 1983).

Up to 1982, Computervision had 75 per cent of the Swedish market, but when CIS came onstream in that year the new company made very substantial inroads into the Computervision market share. This was said to be because the CIS system integrates better into wider computerisation within firms. IBM is likely to do well out of Swedish firms' desire to integrate computerisation, rather than treating the drawing office as an island of automation in the classic style of the American turnkeys.

Industrial policy for electronics has been primarily directed at the computer and telecommunications industries. Swedish strength in computers was generated by SAAB's needs for computers for aircraft, and the company's work on computers for this purpose led to a diversification into general-purpose machines. In 1971, the Stansaab computer company was set up as a joint venture between the Standard Radio and Telephone company, SAAB-Scania and the Swedish Economic Development Agency. In 1978, Stansaab merged with SAAB's own data division to form Datasaab, owned equally by the state and SAAB-Scania. Datasaab's product development has subsequently been supported with state funds.

The principal external source of support for CAD/CAM technology development has been the Technology Development Directorate (STU), which funds research in technology in academia and industry. While it is readily acknowledged that Sweden has neither a sufficiently large supply industry nor a big enough home market to permit the domestic generation of much of the technology needed by the engineering industry, it is not felt that needed technology can simply be bought in from abroad. STU takes the position that Swedish industry has particular technological requirements because of its need to operate in international markets. Equally, imported technology cannot be exploited unless there already exists relevant know-how at home. STU therefore funds two types of work on CAD/CAM, as in many other technologies: a programme to develop technological capability (*Kunskapsuteveckling*); and a separate programme for developing technology (*Teknikomraden*), often involving joint funding of projects with industry.

The Computers and Electronics Commission (DEK) was set up in 1978, and has formulated proposals for industrial policy to promote the use of electronics technology in Swedish industry, including detailed proposals for a high technology CAD/CAM and robotics centre to be attached to each of Sweden's five technical universities (DEK, 1980), with government-allocated funds to be matched by the respective trade organisations. The matching funds, however, did not materialise, and only one centre (at Linköping) was established. A local government initiative subsequently produced a similar centre at the Luleaa technical university.

In 1980, a four-year joint venture in electronics CAD (CAD80) was launched, involving SAAB-Scania, STU and two universities, with SAAB-Scania and STU sharing the project cost of Skr. 10m ($2.4m). SAAB-Scania expects to invest a further Skr. 3m ($0.7m) to market the resulting CAD/CAM system on a bureau basis to small and medium-sized firms. STU has a further R&D agreement for CAD/CAM, robotics and related technologies with the Swedish Association of Mechanical and Electrical Industries covering 1980–85, with the respective partners contributing Skr. 45m and Skr. 48m, (approximately $11m each at 1980 prices) (Computers and Electronics Commission, 1981).

NORWAY

Norway's self-perception as a small developing country has led to an industrial policy focus on diversification away from primary industries into specialised market niches and avoiding competition in mass markets. The government report 'Structural Problems and Growth Opportunities for Norwegian Industry' (Lied, 1979) stressed this point, and argued the need for exploitation of CAD, automation, microprocessors and robots. The current weight in its industrial economy of the North Sea oil sector, the krone's status as a petrocurrency, and the high level of wages and social payments in Norway in recent years have increased these needs.

As in other countries, the Norwegian CAD market is effectively dominated by American suppliers. Computervision is the market leader in turnkey systems. The only domestic producer of any significance is SI—the Central Institute of Industrial Research. This was set up by NTNF (the Norwegian Technology Research Council) after the last War as part of the reconstruction effort and built the first Norwegian computer. Its largest CAD project is called Autokon and is a suite of programs developed for the shipbuilding industry, reflecting Norway's strength in shipping and shipbuilding. Today Autokon dominates the small international market for specialised shipbuilding CAD software.

CAD use is predominantly by big firms, although bureau activity is starting, bringing access to the technology for a broader range of firms, especially in PCB design. In line with changes in the national pattern of industrial activity, the focus of Norwegian

CAD has shifted from shipbuilding to offshore engineering in recent years.

Norway has not had a significant domestic computer manufacturer until recently, when Norsk Data has become relatively successful in the sale of high specification superminicomputers. It is now offering a third party CAD/CAM software packaged into turnkey systems.

Two other well-known groups are also involved in CAD: Den Norske Veritas has sophisticated design analysis software available via its bureau service Computas; and Kongsberg has a joint venture with the British Technology Group (the British government's venture capital group). Kongsberg's strength is in automation (CAM) rather than design, and the company is an important supplier of NC controls and programing systems.

On the electronics side, there have been two important initiatives. ELDAK involved the electronics group at SI and produced a schematics package called CASS, which is marketed as part of the ICAN product. ICAN is a small CAD firm located at Horten, the small town just outside Oslo which is the centre of Norway's small CAD industry, and was started by three people from SI and one from another research institution. The company's strategy is to manufacture sophisticated workstations and to sell software. ICAN exports to other Scandinavian countries, and has established branch offices in Stockholm.

Norway itself provides only a small market, so Norwegian CAD products have to be marketed internationally if they are to generate sufficient sales to keep a vendor afloat. Norwegian CAD strategy has been to operate in tailor-made systems, rather than to meet the big league in head-on competition. SI has been involved in discussions with Computervision about the addition of certain Autokon modules to Computervision's product range. Once the Computervision 32-bit host computer is established in the market, it will be possible for SI and others to sell their software to existing turnkey CAD users, and this may be a feasible strategy for at least some part of the Norwegian CAD industry.

Government policy has been generally supportive. Since the last War, technological and research measures to aid Norwegian industry have been channelled through NTNF into a number of research institutes. Co-operative projects with industrial companies and with partners in Scandinavian and German research institutions have been undertaken in CAD and CAD/CAM. The total budget for CAD was about Nkr. 21m ($3.3m) in 1982. This split three ways:

1. Advanced Production Systems (APS); a co-operative venture with the West German research ministry (BMFT). Norwegian participants are SI and the technological University at Trondheim (NTH/SINTEF).
2. SI/Aker Group (SIAG).
3. Geometriske Produkt Modeller (geometric product models: GPM). This is a university sector co-operation between SI and SINTEF (the Industrial and Technical Research Company, associated with the Technical University in Trondheim) and partners in Sweden and Denmark. In 1983, the annual level of funding was running at about Nkr. 3m per year ($0.41m), and the total effort expended had been about 40 person-years (STU, 1983). The 'Nordisk Fund' funded some of this. The Norwegian partners see themselves as putting a disproportionate amount of know-how into this project, which is then exploited by the other partners.

Until 1983 there was no Norwegian policy on CAD but in 1983 proposals were presented by NTNF to co-ordinate Norwegian policy in this area. This resulted in a liaison committee being established, where representatives of policymaking bodies whose actions affected CAD/CAM in Norway could meet.

Policy implications

European interest in CAD in the 1960s was predominantly in the aerospace, automobile and shipbuilding industries, followed by cartography and with the electronics industry becoming involved by the early 1970s. At that initial stage, CAD addressed problems associated with the capture of design geometry and its subsequent manufacture using numerically controlled machinery of various types. The use of CAD systems as drawing machines largely passed Europe by until the late 1970s, when American firms began their European marketing thrusts. The disparity between the use of CAD systems in the United States of America and in Europe is therefore considerable.

Available data comparing CAD usage across the European countries studied need to be treated with caution for several reasons: their sources and bases are in some cases liable to induce biases; the sectoral distribution of industry varies between countries, leading to national needs for different types and quantities of CAD; and also, CAD is now diffusing rapidly, so that even the best of all possible data collection systems could produce only out-of-date indicators and comparisons.

The available data are collected together in Table 2.7 and used to generate crude measures of diffusion. The importance of CAD in relation to population and gross domestic product (GDP) is greater in Scandinavia than in other places studied. Arguably, this relates to the importance for Norway and Sweden of activity in specialised industrial sectors where the amount of value added in the product needs to be high in order to support high wage levels, and where national competitiveness is substantially founded on non-price factors.

Given the crudity of the data upon which Table 2.7 is based, it is not possible to distinguish meaningfully between the level of use of CAD in West Germany and France, though the use of CAD may involve larger single-site installations in Germany than in other countries. The United Kingdom occupies an intermediate position

Table 2.7 International comparisons of CAD use, 1982*

	1980 Population (000)	1980 GDP (billion ECUs)	1982 CAD Systems	CAD per 1,000 population	CAD per bn ECUs of GDP
Norway	4,086	44	70†	0.017	1.59
Sweden	8,311	88	208	0.025	2.36
France	56,345	470	562‡	0.010	1.20
FR Germany	60,095	587	375‡,§	0.006	0.64
UK	57,027	377	620§	0.011	1.64
EEC-10	270,857	2,017	..	..	..
USA	227,658	1,867	6,600‖	0.029	3.54

* The data sources for CAD systems are crude and, in various ways, biased. The data presented here should, therefore, be treated with caution and scepticism.

† 1983.

‡ In estimating the total number of systems for France and West Germany it has been assumed that IBM's market share of CAD is the same as its share in the world overall.

§ Projected on the basis of 1981 figures plus 50 per cent. World-wide shares of CAD grew 30 per cent in this period, somewhat slower than in recent years. Here it is assumed that European CAD markets were growing faster than the American market, being in an earlier stage of development. The British Department of Industry estimated the British CAD park at 1,000 systems in mid-1983, which is broadly consistent with the arbitrarily assumed growth rate of 50 per cent between 1981 and 1982. Another estimate places the British CAD park in September 1983 at some 1,400 systems, which would imply that the British estimate made in the table above is conservative.

‖ Estimated on the basis of Computervision's stated American base and Merrill Lynch market share estimates.

Source: Statistical Office of the European Communities, Chap. 4, 1983; Kurlak, 1982.

between these countries and Scandinavia (despite the United Kingdom's considerably lower GDP per head than the other countries considered) and is probably the major European user of CAD. The comparative importance of CAD in the United Kingdom relates partly to the strategies of the American CAD suppliers, for which the relative absence of a language barrier makes the United Kingdom the natural first port of call in marketing into Europe. Most of them initially established their European headquarters in the United Kingdom. More recently, there has been a trend towards setting up separate nationally or regionally based divisions within Europe as the market has grown. It is difficult to assess the importance of the British Ministry of Technology's activities during the 1960s, but they clearly played a role in promoting the supply and use of CAD which was unparalleled by policy measures elsewhere in Europe. Recent financial inducements to CAD adopters have also played their part although the CAD market in the United Kingdom was clearly in substantial growth before the announcement of the Department of Industry schemes.

Despite the importance of European activities in certain specialised areas, CAD is largely a US-originated technology and the United States dominates CAD markets in Europe as elsewhere. This dominance is reinforced by American strength in the computer industry, and the corresponding weakness and national orientation of European computer suppliers. Such success as European CAD companies achieve largely involves the use of American hardware.

The larger European countries considered here have a history of government intervention in computers, based on the perception that it is important to have a national presence in that industry. While they have remained important in their own countries, these European suppliers have had only limited international success. Policies to promote CAD diffusion have not been systematically related to other electronics policies, while government industrial policy for the CAD supply side in Europe has been notable chiefly by its absence. In general, European countries have appreciated the need for technological capability in CAD, if only as a basis for effective use of the technology. However, the extent to which this capability is diffused from the higher education and research sector into industry is variable.

CAD poses two principal types of policy problems for European countries: how to adopt the technology effectively, so as to promote national industrial competitiveness; and whether to promote a national presence in the CAD supply industry. In the absence of a co-ordinated approach to these problems at the national level, the effects of policy may be contradictory. Thus, policies designed to promote adoption can strengthen the position of foreign suppliers in the national market, making it very difficult for national supplier companies to enter and survive.

As far as use is concerned, the take-off and use of CAD in European countries has been largely prompted by imports of the American turnkey systems. The origin of these systems as packaged forms of drawing office software dating from the early 1970s necessarily shaped their architecture. It is no great surprise to find the United Kingdom amongst the leaders in the use of these systems—the proneness of British firms to management by profit-centre in the American style makes the American approach more suited to the United Kingdom than to other European countries where the drawing and (creative) design functions are more closely integrated.

The focus on drawing office activities brings productivity improvements in this sphere. But a really important role for CAD is as a stepping stone towards computer-integrated manufacturing (CIM). Increasing use of computers in design and manufacturing and to control information flows, allows companies to be operated more effectively as the integrated systems. In this respect the turnkey systems have their limitations, and it is interesting to observe how some of the more sophisticated Swedish and German users, as they advanced along the learning curve, have sought CAD applications that have extended the capabilities of the systems (e.g. the CIS Medusa package). However, even the turnkey systems have some advantages in this respect. First, they can provide a 'user friendly' introduction to the potential of more sophisticated systems. Secondly, they provide a useful stimulus to changing methods of training. In the absence of adequate supplies of manpower trained to use CAD and other computer-based production technologies, company and national competitiveness can be seriously impaired.

On the supply side, it is interesting to note that unlike the American CAD companies which achieved success in the 1970s (Computervision, Calma, Applicon and so on), European firms actively trying to market CAD in the 1960s (Ferranti, Racal, Aristo, Kongsberg) were already well established. For them, CAD was a relatively minor product diversification, while for the new American firms it was the basis for survival and growth.

One of the interesting contrasts between European and American experience is the failure of these European firms to 'take off' and develop the technology in the early 1970s in the same way as their American counterparts. Today, European CAD suppliers still follow little league, 'niche' strategies, and are unable to compete with the breadth of competence of the American big league firms which grew from smaller bases. One explanation for this, as we have seen, is the failure of the European computer industry to provide the stimulus to the CAD firms to shift to packaged systems based on stand-alone minicomputers. But another explanation is market fragmentation (and this of course also afflicted the computer companies). The European market is fragmented by language differences, standards and national chauvinism. Companies' initial orientations in France, Germany and the United Kingdom tend to be in terms of the national market. Only the Scandinavians are obliged by the small size of their domestic markets to think in international terms from the outset.

It is now difficult to envisage any European firm following a big league, multi-application product strategy on the basis of its own rather restricted national market. The American companies have the advantage of substantial market penetration with established sales and service networks, and all the learning curve advantages that go with this. Yet Europe as a whole—or even the EEC—forms a bigger market than the United States. The sad reality of the situation is that in national policy, as in company strategy, the Europeans react individually to the United States. It is clear, for example, from discussions in France and the United Kingdom that French practitioners regard France as second only to the United States in CAD while British practitioners, correspondingly, regard the United Kingdom as occupying this position.

Apart from academic co-operation involving low levels of funding within Scandinavia and between West Germany and Norway, the co-operative spirit is absent from the CAD area. In the meantime, the American CAD suppliers have been able to pick off some of the finest fruits of European labours with this technology. In France, Battelle and IBM market SIGMA, SYSTRID and CATIA. In the United Kingdom, Computervision owns CIS; Evans and Sutherland own Shape Data, which originated the ROMULUS modeller; Prime owns Compeda, and has partial rights to PDMS.

The case for a European initiative

The minimum policy requirement in European countries is the development and maintenance of a basic level of know-how orientated towards the development and use of CAD technology. This permits some level of effective use both of CAD itself and of other aspects of CIM, and is the strategy followed by West Germany and Sweden, countries which have effectively opted out of the CAD supply industry.

While this minimum level of technological capability is clearly necessary for European nations to remain in effective competition with other industrial countries, the case for state intervention to create a significant capability on the supply side is more ambiguous. The strength of the case rests largely on judgements about the likelihood of alternative political scenarios and about the importance and relative sizes of potential synergies between local, as against American, suppliers and European users.

The early marketing of relatively inexpensive, turnkey CAD technology in the United States has provided American users with earlier access to the technology than their European competitors and allowed them to progress through the relatively protracted learning process associated with CAD adoption ahead of their European counterparts. The American origin of the dominant supplier firms means, also, that the technology is supplied to European users on poorer terms—both with respect to prices and after-sales service—than to American users. Dependence on imported technology, therefore, leaves the European user vulnerable to comparative disadvantages *vis-à-vis* users in the United States. In the case of CAD, these disadvantages are exacerbated—and, in the present political and economic climate may be yet further affected—by American policies designed to inhibit technology transfer.

At the political level, the case is based on the increasing estrangement between the United States and other members of the Western alliance and the growing isolationism of American positions on technical transfer. If Europe is to remain competitive there is need for European capability at the leading edge of technologies such as CAD.

In strict economic terms, on the other hand, it may not matter whether CAD and other technologies are supplied from Europe or from elsewhere, provided best practice is available to users. It is not clear that generating more European activity on the supply side of CAD would provide sufficient benefits to outweigh the costs and delays to potential users associated with the early years of establishing that activity.

Nevertheless, there are some good arguments in favour of trying to create a European capability on the supply side. First, the accumulated weight of high-quality European CAD software is evidence of the existence of European competence which could form the basis of a big league supplier. Anglo-French modelling capability is high. PCB design is also an area of considerable success for these countries.

Pipework capability exists in the United Kingdom (and in Holland), while there is competence in NC programing and control technologies spread right across the continent. There is, to our knowledge, no European competitor to Calma's dominance of VLSI applications in the market for traded CAD systems, but the ability to build corresponding European technology exists in a number of European semiconductor companies. As around MIT in the Boston area of Massachusetts, there are places in Europe where CAD firms congregate in a sort of 'Silicon Valley'. The most important of these is probably Cambridge, England, but a similar effect occurs at Horten in Norway and there is a somewhat more diffuse concentration of supply capability in South Germany, providing some possible geographical growth nodes.

Secondly, CAD, as we have seen, is an 'island of automation' being increasingly absorbed into the more general automation of manufacturing industry—CIM. One argument for promoting a European CAD industry is that it would help create a body of skilled personnel able to promote and smooth the adoption of CIM.

Thirdly, the American turnkey versions of CAD were developed for the American business environment and do not perform as productively outside this environment. It can be argued that the optimal technology to be employed in European design and drawing departments is likely to be one developed in and for Europe.

There are good arguments also for taking an initiative now. The big league firms are under pressure from both the little leaguers —who, like CIS, with its MEDUSA system, can provide a more appropriate package than the big turnkey systems—and the major computer firms like IBM who can offer more comprehensive automation systems. These pressures have already led to an 'unbundling' by the big league suppliers of their systems. (These companies had previously confined use of the software they supplied to their own hardware.) This in itself provides an opportunity for little league firms to extend their activities. But the general reshuffling of market positions that is currently taking place would also provide an opportunity for a new 'European' firm to enter the market.

However, any intervention on the supply side would be both difficult and expensive. Despite the problems which have led them to unbundle their products, the dominant American turnkey suppliers are by no means a spent force—witness their recent acquisition of some of the most promising European little league firms. The increasing role of mainframe computer suppliers—notably IBM— is also evidence of a new type of tough competition in CAD supply. The failure of the attempt by the Department of Industry in the

United Kingdom to promote a union between Racal, Quest and Compeda illustrates that more is involved in a big league supply position than merely having a broad range of products. The difficulties of realigning incompatible systems are great, and this calls for technical acumen in policymakers attempting to intervene in this sector.

The conclusion that one comes to therefore is that there is scope for European entry into the big league, but probably only on an internationally co-operative basis, which would mean national governments within Europe being willing to cast aside their chauvinism and back the European initiative with both finance and political will. History teaches us to be sanguine about the prospects for the type of European co-operation which would be necessary to underpin a substantial intervention in the CAD industry. Although there have been important successes, such as the European Airbus, the use of the ESPRIT advanced information technology project as a political football in EEC debates about the common agricultural policy during 1983 and 1984 is a salutary reminder of the European taste for playing the chauvinist.

Bibliography

Arnold, E. and Senker, P. (1982), *Designing the Future: The Implications of CAD Interactive Graphics for Employment and Skills in the British Engineering Industry*, Occasional Paper No. 9, Watford, Engineering Industry Training Board.

Carlsson, Jan (1982), *Presentation to IBM Workshop on Automation in Manufacturing: Effects on Productivity, Employment and Worklife*, mimeo., Stockholm, Computers and Electronics Commission.

Computers and Electronics Commission (1981), *The Promotion of CAD/CAM and Robotics in Sweden*, Stockholm, Computers and Electronics Commission, Ministry of Industry.

DEK (Data- och Elektronikkommitten) (1980), *Datateknik och Industripolitik*, SOU 1980:17, Stockholm, Statens Offentliga Utredningar.

DEK (Data- och Elektronikkommitten) (1981), *Datateknik i Verkstadsindustrin*, SOU 1981:10, Stockholm, Statens Offentliga Utredningar.

Green, R. Elliot (1969), 'Computer Graphics in the United States', *Computer-Aided Design*, 1(3).

Irvine, John, Martin, Ben, R. and Schwarz, Michiel, with Keith Pavitt and Roy Rothwell (1981), 'Government Support for Industrial Research in Norway: A SPRU Report', Appendix 4 to *Forskning, Teknisk Utvikling og Industriell Innovasjon: En Vurdering av den Offentlige Støtte til Teknisk-industriell Forskning og Utvikling i Norge*, NOU 1981:30B, Oslo, Norges Offentlige Utredninger.

Kaplinsky, Raphael (1982), *Computer-Aided Design: Electronics, Comparative Advantage and Development*, London, Frances Pinter.
Kurlak, T. (1982), *CAD/CAM Review and Outlook*, September, New York, Merrill Lynch.
Kurlak, T. (1983), *CAD/CAM Review and Outlook*, October, New York, Merrill Lynch.
Lewandowski, Steffen (1983), 'Einführung von CAD in der Industriellen Praxis mit Hilfe von Unternehmsberatern', *CAD/CAM Report*, 2/3.
Lied, Finn (1979), *Strukturproblemer og Vekstmuligheter i Norsk Industri*, Oslo, Norges Offentlige Utredninger.
Lindberg, Sten (1983), *Datorstott Konstruktionsarbete—En Kartläggning av CAD/CAM-Tekniken och dess Effekter*, Stockholm, Svenska Industritjänstemannaforbundet.
Nora, Simon and Minc, Alain (1978), *L'informatisation de la Société: Rapport à M. le Président de la République*, Paris, La Documentation Française.
Poitou, Jean-Pierre (1983), 'CAO, réactions des utilisateurs', *Le Nouvel Automatisme*, June, pp. 56–63.
Prais, S. (1981), 'Vocational Qualifications of the Labour Force in Britain and Germany', *National Institute Economic Review*, November.
Riehm, U. and Wingert, B. (1983). 'Technology Induced Morphogenesis of Skills: The Case of CAD', mimeo, Karlsruhe, Kernforschungszentrum.
Roth, Siegfried (1982), 'CAD/CAM in der Automobilindustrie, I–IV', *Angestellten-Magazin*, May, Frankfurt, IG Metall.
Sciberras, Edmond (1977), *Multinational Electronics Companies and National Economic Policies*, Greenwich, Conn., JAI Press.
Statistical Office of the European Committees (1983), *Basic Statistics of the Community*, Luxembourg.
STU (1983), *Long-Term Development of Competence in Engineering—our Guarantee of Progress*, Stockholm.
Sutherland, I. E. (1963), *SKETCHPAD: A Man–Machine Graphical Communication System*; reprinted in Boston, Garland, 1980.
Wingert, B., Duus, W., Rader, M. and Reihm, U. (1982), *Ergebnisse von CAD-Anwendern im Maschinenbau*, Karlsruhe, Abteilung für Angewandte Systemanalyse, Kernforschungszentrum.

3 Advanced machine tools: production, diffusion and trade

Ernst-Jürgen Horn, Henning Klodt and
Christopher Saunders

The application of microelectronics has become the most important feature of technical advance in manufacturing products and processes. Computerised control of machines, machining centres, even of entire plants or assembly lines, is increasingly common in almost all lines of manufacturing. Modern machine tools (connected with electronics) and industrial robotics (which have evolved from converging developments in machine tools, manual manipulators and remote manipulators) are key areas of change. The purpose of this chapter is to look at what is happening in the industry which in many senses lies at the root of these changes—the machine tool industry. Specifically it is concerned with the development and diffusion of what might be termed advanced machine tools—numerically controlled (NC) and computer numerical control (CNC) machine tools[1] and robotics.[2] Mastering electronics has become increasingly important for success and failure of machine tool producers and users, and for the entry of new suppliers.

Superficially, it would appear that Japan and the United States have gained a lead in the development, production and diffusion of advanced computer control for machinery during the past decade or so. In particular, the Japanese export thrust of recent years, mainly in CNC (computer numerical control) lathes and machining centres, suggests that Western European machine builders have not been able to keep up with their competitors, at least in some technically advanced segments of the machine tools market. If the perception of gaps in technology is correct, the question arises as to whether these gaps will be only temporary or lasting. The popular fear behind this question is that of being overtaken by foreigners over a whole range of prospective growth industries. A much more real possibility is, however, that a country is temporarily falling behind because it is not able to take up the opportunities of technical advance in time.

Governments in Western Europe (and the European Commission) seem to assume that in Europe private firms are not sufficiently dynamic to catch up and to keep track of Japanese and American competitors. Hence the measures taken in various countries to promote development, production and application of advanced machine tools and industrial robotics. In some cases these measures are targeted at these specific areas; in other cases they are less selective and included in more general promotion schemes for information technologies and the like. The reasons for intervention seem to be that the technology concerned is deemed to be a 'generic' one, essential for breeding manufacturing technologies elsewhere, or that machine tools are regarded as a 'strategic' industry occupying a crucial role in the diffusion of technical advance for a considerable part of manufacturing industry.[3] For instance, Rosenberg (1963) argued that the machine tool industry played a key role in the virtuous circle in the industrialisation of the United States. Given the traditionally strong trade protection of the United States in those times, this argument seems to be very much '*post hoc ergo propter hoc*'. Nobody knows what would have happened otherwise.

Later in this chapter we shall be considering the development of government policies towards this industry. Initially, we shall be looking at the recent development of the industry itself, and the emergence of a substantial challenge from Japan in the sector of these advanced, computer controlled machine tools, at patterns of diffusion and the extent to which different countries have adjusted to the Japanese challenge. This is further explored by a detailed look at the EEC trade statistics for recent years, which confirm the picture of Western Europe, particularly Italy and West Germany, regaining some ground despite the incursions of the Japanese. The final section considers the lessons which emerge from this analysis, and particularly the role of public policy in the adjustment process.

Recent developments in machine tools

THE TECHNOLOGY

The range of products offered by the machine tool industry is extremely diversified. Two major distinctions are usually made: that between metal-cutting and metal-forming machines, and that between general-purpose and special-purpose machines. The first numerical control (NC) for metal processing machines was developed in 1952 at the Massachusetts Institute of Technology. NC involved feeding a successive stream of co-ordinates into a machine, usually via punched paper tape, which then guided the cutting tool through these points. It provided faster and more perfect copies than the templates which had traditionally been used. MIT's experiments were of interest to the Department of Defense and became the backbone of their Manufacturing Technology (Man Tech) programme centred on MIT but involving in particular the aerospace industries. However, it was not until 1960 that NC control began to be used commercially for milling and boring and, even so, mainly in the aerospace industry. In the United States, Cincinnati Milacron and Bunker Ramo were the main NC suppliers at that time; in the United Kingdom, where the aerospace industry took an early interest in developments, Ferranti was the main supplier of the control equipment.

The first computer numerical controls (CNC) appeared in the early 1970s. With CNC, direct input of numerical data via punched tape is replaced by the minicomputer, which controls and programs the machine tool. The level of sophistication of CNC machine tools today varies considerably: from simple machine with one or few tools to complex machining centres with magazines capable of holding tools in large numbers. At the highly sophisticated end of the spectrum, computer-assisted manufacturing and flexible manufacturing systems interface with industrial robots. According to a widely accepted definition (International Standards Organisation), 'the industrial robot is an automatic position-controlled reprogrammable multifunctional manipulator having several axes capable of handling materials, parts, tools or specialised devices through variable programmed operations for the performance of a variety of tasks'.

The advantages of CNC machine tools over earlier generations of NC machine tools are listed as follows by Rempp (1982):

— lower hardware costs;
— greater workpiece variety;
— higher flexibility;
— superior quality;
— greater working potential.

Electronic program and control components have become much less vulnerable to dirt, heat and other unfavourable conditions at the workplace. The enormous decrease in the cost of control devices has opened up a range of new applications for CNC machine tools and has made their use quite viable for small as well as for large firms. Today in West Germany, for example, about 50 per cent of these machine tools are used in firms with less than 250 employees. Still more important, the decreasing costs of control have opened up a wide range of new applications. Originally, the technology was mainly used for drilling, milling, polishing and turning; later on it was extended to grinding, punching, forging, torch cutting, plasma cutting, laser cutting and reshaping. Today nearly every metal cutting or forming operation can be handled by these machines.

THE INDUSTRY

The (metal working) machine tool industry has recently experienced a number of distinct changes. First, the links with microelectronic technology have been becoming increasingly close: the emergence of CNC machine tools is a prominent example. Second, the interface of machine tool production with industrial robotics and other forms of computer aided manufacture (CAM), with computer aided design (CAD—see Chapter 2), and recently the advanced flexible manufacturing systems (FMS), has been growing. Third, new sources of supply, in particular Japanese firms, have disturbed traditional structures of international trade and production. Finally, the nature of the industry is changing. Having been formerly a vertically integrated industry, the machine tools industry is, through the link with microelectronics, increasingly becoming an assembly-type industry, importing strategic components, such as electronic controls. It is interesting to note that the suppliers of electronic controls have been reluctant to enter machine tool production proper, although they have moved into industrial robotics. The latter is at the most sophisticated end of the spectrum where microelectronics counts most. Potential users, particularly automobile firms, have pioneered many technical developments in this area.

The machine tool industry as a whole is an old industry and one of the main producers of capital equipment. It is a relatively small one in relation to Gross Domestic Product (GDP). For instance, in West Germany sales of the industry amount to 0.6 per cent of GDP, and in the United States—the world's biggest producer of machine tools—to 0.2 per cent of GDP. The industry is somewhat unevenly distributed across countries. Table 3.1 shows the shares in world production and exports of the world's leading producers. It shows that there is no close link between machine tool production and the international pattern of overall economic performance by country (which sheds some doubt on the hypothesis of the strategic role of this industry). Table 3.2 shows

Table 3.1 Shares in world machine tool production*
and exports, 1981 (%)

	Production	(of which NC)	Exports
US	19.5	(39)	11.1
W. Germany	15.3	(20)	24.2
Japan	18.4	(50)	16.4
UK	3.5	(19)	4.7
France	3.1	n.a.	3.8
Italy	5.3	(22)	6.8
Switzerland	3.1	n.a.	7.0

* In value terms.
Source: Ray (1984) Table 7.2 and Chart 7.3

Table 3.2 Shares in world exports of machine tools,
1965–1981 (%)

	1965	1975	1977	1981
US	22	12	9	11.1
W. Germany	31	36	35	24.2
UK	13	8	6	4.7
Japan	3	7	10	16.4

Source: Daly and Jones (1980); Ray (1984).

export shares since 1965 and illustrates the dramatic rise of the Japanese industry.

The machine tool industry is not a concentrated industry. In most countries it consists of a small number of big firms, most of which already played a distinguished role in the industry at the beginning of the century, and a large number of middle-size and small firms. This is illustrated by Table 3.3. The larger firms normally meet the demand of concentrated sectors such as automobiles, aerospace and armaments, while small firms are highly specialised in many different market segments with much smaller-scale demand. Looking across countries, it appears that the Japanese machine tool industry, a newcomer by international standards, is the most concentrated, but only in specific sectors of the market. Overall, the industry is less concentrated than at first sight appears. The reason seems to lie in the fact that in the supply of concentrated sectors a number of

Table 3.3 Size of machine tool firms (1979)

Employment ranges	France		West Germany		Japan	Italy	UK	US
	No. of firms	% emp.	No. of firms	% emp.	% emp.	% emp.	% emp.	% emp.
0–50	105	10.9	160	4.8	11.7	24.0	12.0	15.3
51–100	20	7.2	90	7.5	6.2	16.0	6.0	8.0
101–250	23	18.6	95	18.3	16.2	27.0	11.0	16.3
251–500	12	18.3	55	21.8		13.0	28.0	16.2
501–1,000	7	20.2	35	25.1	13.3	10.0	21.0	23.0
>1,000	3	24.8	15	22.5	52.5	10.0	22.0	21.1
Total	170	100.0	450	100.0	100.0	100.0	100.0	100.0

Source: Rendeiro (1984).

the leading machine tools firms are affiliates of the conglomerates in automobiles, engineering and electronics.

Industrial structure is somewhat different in the most advanced segment of the market—industrial robotics. Here there has been a significant market entry, on the one hand by large companies producing microelectronic control components, and on the other by users of industrial robots such as automobile companies. In both cases, however, production has so far been mainly destined for in-house use. The commercialisation of such developments (domestically or internationally) is mainly left by the large corporations to affiliates or to smaller machine tool firms with whom they co-operate through licensing deals.

The extent of forward and backward integration into robotics by the large engineering and electrical groups is illustrated in Table 3.4. If there had been suitable robots in the market, for instance, the Volkswagenwerk would perhaps never have started the production of these machines, which it needed for the manufacture of automobiles. Today this company is a major producer of industrial robots in West Germany.

In addition to backward penetration there is a strong forward penetration of the market for industrial robots by suppliers of electronics. Nor is it correct that only big enterprises are able to produce industrial robots. In recent years many small firms have entered this market. Some of them (like small enterprises in other markets), have gone into liquidation, but others have survived. In Japan, for example, some one hundred firms are engaged in robot production—about half of them being small or medium-sized businesses (OECD, 1983a).

New capital goods affect the shape of user-industries as well as that of the industries where the new technologies are developed. Traditionally, the products of the metalworking industry were produced either one-off or in small batches, special consideration being given to the needs of the customers. Nowadays microelectronics has given a new flexibility to production—a variety of products can be produced with one machine and all that is changed is the software, not the hardware. This favours the production of hardware in greater batch sizes. Even so, the machine tool industry is not expected to become 'footloose': the knowledge of industrial materials and of mechanical techniques is still too important for the production of machine tools to become a candidate for offshore assembly. Nevertheless, as a consequence of the introduction of microelectronics, the realisation of new economies of scale could be of increasing importance in the future. Recent successes of

Table 3.4 Industrial origin of major producers of industrial robots

Industry of main activity	Firm	Country
Electric and electronic equipment	Yaskawa Electric	Japan
	Hitachi	Japan
	Fuji Electric	Japan
	ASEA	Sweden
	Electrolux	Sweden
	IBM	United States
	General Electric	United States
	Texas Instruments	United States
	Olivetti	Italy
	Siemens	West Germany
Machinery	Unimation	United States
	Cincinnati Milacron	United States
	Westinghouse	United States
	Kuka	West Germany
	Fanuc	Japan
	Trallfa	Norway
	Rosenlev	Finland
Transport equipment	Kawasaki	Japan
	Kobe Steel	Japan
	Volkswagenwerk	West Germany
	General Motors	United States
	Comau-Fiat	Italy
	Renault	France

Sources: Schenk (1982); OECD (1982); Batelle Institute.

Japanese and Italian firms, for instance, seem to be largely based on the fact that they first realised the scope for larger-scale production of the multi-purpose machines which were developed as a result of progress in control devices. An interesting question is how far these opportunities for scale economies will reach and what effect they will have on the structure of the industry.

Perhaps the most important change has been the entry into the machine tool industry of the specialist suppliers of electronic equipment. There are a few 'big league' machine tool companies which have entered the manufacture of electronic equipment (for example Cincinatti Milacron, Gildemeister, and the 'old' Alfred Herbert), but these are really exceptions. On the whole, it is the traditional electrical and electronic companies (for example GE, Fanuc, Siemens,

Allen Bradley, Bosch, Olivetti, Philips) which have become the main suppliers of NC and CNC controls. Interestingly, most electronic suppliers have not sought to enter the traditional core of machine tool manufacture—they remain component suppliers outside the main industry proper. Only in a few cases (for example Fanuc, Olivetti) is their dominant interest in manufacturing automation and robotics.

The development of NC machine tools has also led to vertical disintegration in another section of the industry. The traditional machinery maker was a vertically integrated company, particularly in relation to manufacturing. In an environment where craftsmanship and mechanical technologies predominated, making (rather than buying in) parts and components was the way to ensure control over the overall performance of the machine. Today, all that is required is for the manufacturer's specifications to be embodied in the program controlling the machining centre. This can be done in-house or by the component supplier, but the trend towards using subcontractors has become stronger in the last few years as controls have got more and more sophisticated. In this sense, machinery companies are becoming more of a design, assembly and marketing operation, rather than a manufacturing concern. The adjustment pressure for these companies (particularly the major producers) has thus been really how to move from a complete emphasis on mechanical technologies and manufacturing, into a new emphasis on R&D and marketing.

In the course of the last decade two important developments have emerged. The first is the supercession of NC machine tools by CNC machine tools. For instance, in West Germany, in 1977, the share of CNC machine tools in total NC machine tools production was only 33 per cent. By 1982, this share increased to 96 per cent. The second major development has been the 'take-off' of robotics which came a little later, in the early 1980s. The 'Verband Deutscher Maschinen und Anlagenbau' (VDMA) estimated that 7,500 robots would be in use in West Germany by 1985. Annual production surpassed 2,000 units in 1983.

The use of industrial robots has been largely the domain of the big engineering groups, in particular automobile manufacturers. In the production of industrial robots, a more differentiated picture has emerged than for CNC machine tools. On the one hand, large car manufacturers and electronics firms have taken up development and production of robots, initially often for in-house use. On the other hand, a number of relatively small firms have begun specialising in the production of robots (Junne and van Tulder, 1984). It seems likely that it will become increasingly efficient for smaller firms in the future to use robots as the range of applications is extended and the robot technology down-sized to meet the needs of smaller firms.

Diffusion patterns

Industrial robots and CNC machine tools are just entering a period of rapid diffusion in industrialised countries. These technologies have been on the market for about ten years now. The acceleration of growth rates begun in the second half of the seventies has not yet come to an end. Figure 3.1 gives a simplified diagram of a 'normal' diffusion pattern, and of the stage which industrial robots and CNC machine tools seem to have reached. The application of CNC machine tools is somewhat more advanced than that of industrial robots; CNC machine tools entered the market a few years earlier. But CNC machine tools, as well as industrial robots, are obviously facing a potentially rich market in the future. Special diagrams

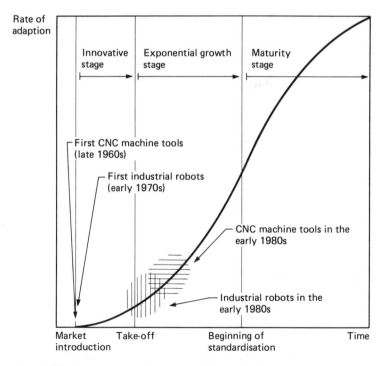

Figure 3.1 Diffusion patterns of CNC machine tools and industrial robots.

might have been drawn for various types of these machines, since there are great differences between more or less intelligent versions. In the years ahead, new and better machines will be constructed, while older ones will be standardised or become obsolete. This is indicated in the diagram by the range of possible locations of industrial robots and CNC machine tools at the present time.

DIFFUSION OF CNC MACHINE TOOLS

Machine tools was formerly an industry where production was concentrated on few countries. International trade intensity was very high; international investment and co-operation was low; and access to licences from foreign producers was relatively easy. Here again things have changed substantially with the emergence of microelectronics. The internationalisation of machine tools firms via foreign investment and co-operation agreements has increased considerably. The larger firms have been particularly active in this field. Three factors seem to be important. First, Japanese firms know that they may face increasing import restrictions by other industrial countries. Secondly, the trend toward standardisation (general purpose 'catalogue-machines') has increased the scope for splitting internationally, design, assembly, component manufacture and marketing activities. Production and subcontracting can be easily relocated internationally. Third, indivisibilities in technology transfer have been much reduced. The machine tool industry has thus become much less national in character.

The post-war history of international competition in machine tools provides clear evidence that being the first in innovation hardly ever leads to the ability to out-compete foreign producers in the medium or longer run. One striking example is the emergence of NC machine tools since the late 1950s. Table 3.5 shows how strong a lead the United States had over Europe in the late 1960s, and how, at that time, the United Kingdom was well ahead of its European competitors. Japanese suppliers were hardly known then on Western markets. The American firms could not, however, maintain their original lead in commercial applications, and by 1980/81 both the Europeans, and even more so the Japanese, had closed the gap very considerably.

Special data on CNC machine tools are available for a few countries only. In international comparisons, one has to rely on data on NC machine tools, which include CNC machine tools. Increasingly, there is no distinction between NC and CNC machine tools. As we have seen, in West Germany, for example, in 1982 96 per cent of

Table 3.5 Stock of NC machine tools, 1969 to 1980/81 (numbers of machines '000)

	1969	1975/6	1980/1
US	20.0	40.0	80.0*
W. Germany	1.9	8.0	25.0
UK	3.2	9.7	25.8
Japan	n.a.	14.0	28.0*
France	n.a.	4.0	16.4
Italy	0.8	3.3	11.4
Sweden	0.5	2.1	4.0

* Estimates derived from qualitative data given in National Academy of Engineering (1983).

Source: Ray (1984) except starred items.

NC machine tools in use were in fact CNC machine tools. Given this, and the fact that from this point on we are concerned mainly with international comparisons, we shall for the rest of this chapter make no distinction between NC and CNC machine tools. NC implicitly means CNC.

Table 3.6 extends the information given in Table 3.1. It shows that for NC machine tools, the major producing countries are Japan, the United States and West Germany, and that by the second half of the 1970s Japan had overtaken the United States and West Germany. But it has not yet overtaken the European Community as a whole.

Table 3.6 Production and final consumption of NC machine tools (values in $m), 1976 and 1980

Country	Production		Final consumption	
	1976	1980	1976	1980
Japan	180	1,560	270	780
United States	510	1,250	430	1,460
European Community	550	1,820	340	1,560
W. Germany	270	920	150	750
Others	110	570	310	1,400
World	1,350	5,200	1,350	5,200

Sources: Fachgemeinschaft Montage, Handhabung, Industrieroboter im VDMA, unpublished data. Commission of the European Communities, The European Machine Tool Industry (1983).

The difference between production and domestic investment (final consumption) yields net exports of the country concerned. In the relationships between these figures, remarkable changes have taken place. The United States was and is the country where most NC machine tools are used; but in recent years the United States has become a net importer, and its importance as a producer has declined considerably. On the other side, Japanese net exports equalled more than half of production in 1980. Only four years before, net imports had equalled half of production (and one-third of final consumption). The dramatic growth of Japanese NC machine tool production was export-led; final consumption did not grow more rapidly in Japan than in the United States or the European Community. The latter (and among its member countries the most important producer, West Germany) has remained a net exporter in NC machine tools, but in relative terms the foreign trade position has worsened in recent years.

The figures for final consumption show that the greatest increase (measured in terms of number) in the installation of NC machine tools has taken place in the European Community. By 1981 the number of NC machine tools installed in the Community as a whole was greater than in Japan and equalled the number in the United States (Table 3.7). The percentage of the total machine tool park taken by NC machine tools does not differ very much among the European countries listed in Table 3.7. The United Kingdom and Sweden show the highest, and France the lowest density of NC machine tools.[4] It is remarkable that the United States has now been

Table 3.7 NC machine tools as percentage of total machine tool park

	Year	Stock '000s	% of total MT park
US	1976	40.0	1.5*
West Germany	1980	25.0	2.2
UK	1981	25.8	2.5
Japan	1976	14.0	1.5*
France	1981	16.4	2.0
Italy	1981	11.4	2.2
Sweden	1980	4.0	3.0

* These figures relate to 1976, whereas the European figures relate to 1980 or 1981. Given that both the United States and Japan approximately doubled their stock of NC machine tools during those five years (see Table 3.5), it seems likely that in both countries usage now approaches the 3 per cent level.
Source: Ray (1984), Table 7.4 and Chart 7.5.

caught up by Japan. The technological leadership of the United States, where NC and CNC machine tools had been originally developed, has been steadily eroded since the mid-1970s.

Diffusion of robots

There are various statistical definitions of industrial robots, which differ from institution to institution and from country to country. This raises further problems. Here we will define industrial robots as handling or working machines which are freely programmable, with three or more motion axes. Simple handling manipulators or tele-operators are thus excluded. This is a relatively narrow definition, which is used in most European countries.[5]

According to this definition, the United States was the main producer of industrial robots in 1974 (Table 3.8). Seven years later, Japan was producing more industrial robots than the United States

Table 3.8 Production of industrial robots* (units)

Country	1974	1980	1981	1982	1985
Japan	400	3000	4000	•	10000
United States	600	1850	2100	•	8000
Europe	400	1500	1700	2500	6300
West Germany	•	800	1200	1600	3200
Sweden	•	•	300	•	•
Italy	•	•	260	•	•
United Kingdom	•	•	200	•	•
France	•	•	100	•	•
Switzerland	•	•	30	•	•
Austria	•	•	20	•	•

* Estimates according to common definitions for all countries. See text.
Sources: Fachgemeinschaft Montage, Handhabung, Industieroboter im VDMA, unpublished data: Wolfsteiner (1983); own calculations.

and Europe together, and has now become the most important supplier all over the world. It seems likely, however, that Japan's current advantage will decline in the years ahead.

Among the nine dominant producing enterprises within the OECD area, five are Japanese firms (Table 3.9). There are only two firms from European countries. It should be noted, however, that the largest German producer of industrial robots, the Volkswagenwerk, does not appear in Table 3.9; it utilises most of the

Table 3.9 Main producers of industrial robots, 1981

Firm	Country	Turnover 1981 ($m)
Unimation	United States	1197
Cincinnati Milacron	United States	941
ASEA	Sweden	595
Kawasaki Heavy Ind.	Japan	499
Yaskawa Electric	Japan	333
Hitachi	Japan	333
Kuka	West Germany	246
Kobe Steel	Japan	233
Fuji Electric	Japan	200

Source: Schenk and Winfried (1982).

robots it produces in-house in the production of automobiles. Renault (France), Comau-Fiat and Olivetti (both Italy), Electrolux (Sweden), Trallfa (Norway) and Rosenlev (Finland) are other major European producers. In the United States, General Electric and Texas Instruments have started the production of industrial robots in recent years. Together with IBM, General Motors and Westinghouse, they can be expected to expand their market share quickly and the importance of American suppliers will probably increase in the future.

Japanese enterprises are currently the main users of industrial robots (Table 3.10), but their dominance is less pronounced than in production, since the export-intensity in industrial robots of Japanese producers is higher than in other countries. Nevertheless, in Japan, the number of robots installed per person employed is higher than elsewhere, except in Sweden. In the European Community, West Germany has the highest density of industrial robots in use, followed by France.

All in all, it must be recognised that Japanese enterprises have made best use of the opportunities offered. They have not only gained a strong position in export markets for industrial robots, but they have also been relatively successful in renewing their own capital stock. This gives further advantage to Japanese enterprises in a variety of industries, for example in the production of automobiles. On the other hand, the rapid diffusion of robots in Japan has provided a strong domestic base for producers of robots, which in turn helps cut costs for export markets.

Table 3.10 Industrial robots installed (units), 1974-1985

Country	1974	1980	1981	1982	1985*
Japan	1,500	6,000	9,500	13,000	25,000
United States	1,200	4,500	7,000	9,000	20,000
Europe	500	4,500	6,500	9,000	15,000
West Germany	130	1,200	2,300	3,500	7,500
Sweden	140	1,100	1,700	2,000	2,800
Italy	90	400	450	800	•
United Kingdom	50	370	700	1,200	•
France	30	580	800	1,000	•

* Forecasts
Sources: Fachgemeinschaft Montage, Handhabung, Industrieroboter im VDMA. Unpublished data: OECD (1982); own calculations.

Country-based developments

The following section briefly traces the development of the machine tool industries of West Germany, Italy, the United Kingdom and France during the last decade, concentrating on competition in NC machine tools, particularly from Japan, and on the role of government in seeking to promote and develop indigenous industry in the face of competition.

WEST GERMANY

The West German machine tool industry is the second largest producer in the world and the largest exporter (see Table 3.1). Its emphasis has always been upon high quality, high performance machines which have sold on name and reputation rather than price. It is made up of a large number of small firms (only 15 out of 450 have more than 1,000 employees—see Table 3.3), with median employment slightly below 500. A great many firms are family businesses which have tended to specialise within relatively narrow market segments. This diversity has in the past given the industry great flexibility, but with the advent of the general purpose CNC machining centres, adjustment has proved more difficult because, as pointed out earlier in this chapter, it has turned upside down the traditional economics of the industry. There are now benefits to be gained from relatively long (300-400) runs of standard machining centres or CNC lathes, whereas the flexibility offered by these machine tools reduces the advantages of standard runs and large batch production in the engineering industry using these tools.

The concentration on the high-quality, customised product helps to explain the rapid penetration of the German market by Japanese NC machine tools when they first appeared in the mid-1970s. The Japanese share of the West German market increased from 5 per cent in 1977 to 13 per cent in 1981, but this included over 50 per cent of the machining centres bought in the country.[6] At the same time, and perhaps more significantly, West Germany's share of machine tool exports slumped from 35 per cent to 24 per cent. This decline was largely attributable to West Germany's failure to match Japan's progress in NC machine tools. For example, in 1980, West Germany's sales of NC controlled lathes were only half those of Japan. But, while markets for NC machine tools were increasing rapidly, those for non-NC tools were stagnant or declining.

The Japanese producers had cost advantages over their German counterparts in terms of longer production runs and the lower cost of electronic controls. It was the larger companies—firms such as Trumpf, Scharmann and Gildmeister—who led the counter-attack. All have undergone radical restructuring, introducing highly auomated production processes into their own plants and, while aiming to retain their reputation for high quality and performance, have diversified their product ranges and, above all, moved into series production of modern CNC machines. The more far-sighted of them have also recognised that CNC machine tools and lathes are but a step towards the more integrated automation of the factory process. Both Trumpf and Gildmeister, for example, have established divisions selling automated systems and have seen sales increase markedly in the last two years.

The Federal Government has not given any sector-specific aid to the machine tool industry, even though many firms have been in difficulty and the banks have had to write off losses for some firms. Nevertheless, both machine tools and robotics are strongly favoured by the more general programme of support for the electronics industry and for the application of electronics to industry. These programmes provided funds of approximately DM40m ($20m) per annum since the mid-1970s, and have recently (1983) been expanded to approximately DM100m per annum ($40m). Furthermore, given an industrial structure dominated by small and medium-sized firms, machine tools seem to have been strongly favoured by the programme of grants for the R&D staff costs for small firms (with a ceiling of DM300,000 a year per firm introduced in 1979). Government assistance under this scheme amounted to DM390m in 1982.[7]

Another important West German research arrangement is the link between Government, universities, the Association of Industrial Research Groups (AIF) and the German Machine Tool Manufacturers' Association (NDW). This collaborative arrangement is directed towards production orientated research projects and government support pays for one-third of the research, with industry paying the balance. Joint industry-government funded institutes for technical training and manufacturing research are attached to many universities. Of particular importance is a consortium of four universities —Berlin, Stuttgart, Aachen and Hanover, which undertake a variety of research programmes as co-operative ventures. Such consortia not only have responsibility for the development of new technology, but also for its diffusion through a wide range of industries.[8]

ITALY

The Italian machine tool industry coped better than most of its competitors with the rigours of prolonged recession. Table 3.1 revealed its relatively high share of world production and exports, with particular strength in the NC markets. This trend has continued and, by 1984, Italy has replaced the United States as the third largest exporter in the world.[9] Much of this growth has been based on its strength in NC and factory automation systems.

The successful development of the Italian machine tool industry goes back to the 1960s and the co-ordinating role played by UCIMU, the trade association for machine tool manufacturers. Recognising at an early stage that the fragmentation of the Italian engineering industry into small, often family-run, concerns put a premium on cheap and efficient credit facilities, it successfully lobbied for tax and soft credit concessions on both home and export sales, and controlled these through two subsidiaries (of the trade association). This enabled it to channel both sales and benefits to its own members, but it also kept it in touch with the market and made it aware of the potential among these small businesses for the all-in-one machining centre. One of the objectives it posed for another initiative—its collaborative R&D programme on advanced machine technology—was the development of a simple, low-cost machining centre for small firms. It was these initiatives of the 1960s (continued through the 1970s) which laid the foundation of Italy's current strength in the automation sector (Rendeiro, 1984). It also provided the Italian industry with a valuable element of protection, particularly against the Japanese.

Italy now has five groups of varying size which combine machine tool and electronics strength. The three leaders are Olivetti, Mandell and Comau, each with the strengths in different areas but currently converging on the broad field of factory automation. Olivetti, well-known for its involvement in office automation, is a leading supplier of CNC control systems, a position enhanced by its link with the American leader in this field, Allen Bradley, which gives Olivetti the right to manufacture and market A-B controls in Europe. Its acquisition in 1981 of Pontigga, the leading Italian producer of CNC lathes, gave Olivetti a foothold in the machine tool industry proper, unusual for a firm whose primary interest is electronics. Mandell is Italy's largest manufacturer of machining centres, being one of the pioneers of Italy's push into this market. Well over 50 per cent of its output is exported and it even has an argreement with Amada of Japan for the manufacture and sale of its models in that market. But it is moving rapidly from machining centres into factory automation and today well over half its revenues are derived from this source.[10]

Italy's best-known company in this area is Comau, a Fiat subsidiary formed in 1977 from three well-established Italian machine tool and materials handling companies. Well known for its work in robotics, the company in fact handles most areas of advanced manufacturing equipment—machining, materials handling, assembly and control. Roughly half its turnover is derived from sales of machining and mechanical assembly systems, and two-thirds of its sales are to the automobile industry. Anxious to diversify away from this dependence, it formed an abortive joint marketing venture with Bendix in 1983, but this was subsequently dissolved in the take-over of Bendix a year later. Nevertheless, the move gave Comau a strong marketing lead into the American market which it is continuing to exploit.[11]

THE UNITED KINGDOM

The British machine tool industry has been decimated by the two recessions of the 1970s and 1980s. Sales between 1975 and 1983 fell by 55 per cent: its share of world exports (13 per cent in 1965) fell from 8 per cent in 1975 to 4 per cent in 1983, while over the same period import penetration rose from 25 per cent to 40 per cent. Japanese penetration of the British market showed a particularly fast increase. For CNC equipment as a whole, the Japanese share of the market rose from 18 per cent (1975) to 37 per cent (1983); for machining centres, import penetration rose to over 50 per cent in 1982, when voluntary export restraint was imposed. (The share has subsequently fallen to 28 per cent.)[12]

Concern about the state of the machine tool industry in the United Kingdom dates back at least a century and there has been a plethora of official reports and committees of enquiry. As long ago as the 1960s, its weakness was recognised to be too much concentration on standard machines and too little on more specialised, high-performance equipment, a trait that became more apparent as time wore on and and export shares diminished.[13] Government policy in the 1960s, however, concentrated on attempts to get more product rationalisation and larger units of production. Alfred Herbert, one of the best-known names in British machine tools, with encouragement and finance from the government-sponsored Industrial Reorganisation Corporation, took over a number of other large machine toolmakers to create a group with (peak) employment of over 15,000. Unfortunately, the fortunes of the group did not prosper; a joint venture with the American Ingersoll group to develop an early form of factory automation proved disastrous, profits declined and by 1974 the group was on the brink of bankruptcy. It was rescued by the Government-owned National Enterprise Board (NEB), with an equity injection of £26m ($60m), and further injections as time went by and trading profits failed to improve. By 1979, with employment fallen to 5,500, it was split into four separate operating groups, with a separate design department. This attempt at reorganisation failed and in 1980 the whole group went into receivership and the four subsidiaries were sold off separately. The name survives today as part of the Tube Investments (TI) Group.

Apart from involvement in the Alfred Herbert Group, the focus of Government policy has been on the demand rather than the supply side of the industry. In the 1960s, a 'pre-production order scheme' involved Government purchase of various 'technologically advanced' (and therefore expensive) machines which were lent free to industrial users; another scheme guaranteed re-purchase of NC machine tools which buyers found unsatisfactory, the idea being to encourage experiment with NC methods. In the 1970s, a scheme under the 1972 Industry Act provided a 25 per cent subsidy on costs of developing and launching new machine tools, and 15–20 per cent grants for projects to modernise and expand capacity.[14] The most recent scheme—the Small Engineering Firms Investment Scheme (SEFIS)—introduced in 1982, again provides grants ($33\frac{1}{3}$ per cent) towards the cost of investment in advanced machine tools. The annual cost of this has been running at approximately £20m p.a. ($30m). These grants apply to British and foreign equipment alike and 46 per cent of expenditures have gone on foreign equipment.[15]

The message implicit in these policies, that British engineering firms badly need to update design and production methods, seems belatedly to have penetrated. A comprehensive survey of metalworking firms in 1982 showed the number of NC machine tools in use to have risen sharply between 1976 and 1982, with Britain now well up amongst the leaders in terms of the percentage of the total stock of machine tools which are numerically controlled.[16] A detailed comparison carried out at the same time between a matched sample of metalworking plants in Britain and West Germany found that West German plants were somewhat more likely to have NC machinery, but that the major difference was in working efficiency. Because of the low technical competence of the British work-force, sophisticated computer controlled machinery in Britain was seldom used to its full potential and was frequently out of commission, waiting for repairs.[17]

NC machine tools is in fact another example of an industrial activity where Britain was in the van of early development but failed to translate its early advantage into wider-scale use and development. Thanks largely to the interest and use by the aerospace industry, in the 1960s Britain came second only to the United States in the production and stock of numerically controlled machine tools. (See also Table 3.5.) But with the relative decline of the aerospace industry in the late 1960s, domestic demand from other industries was insufficient to sustain the momentum and the industry failed to develop export markets.[18] By the late 1970s, the industry found itself totally outclassed by the American, West German and Japanese industries, as witnessed by the import penetration figures quoted above.

Indeed, by 1982-3, the British machine toolmakers were in deep trouble. Markets for traditional non-NC machine tools had all but disappeared. British manufacturers could not compete with the prices being quoted by competitors from newly industrialising countries. The message that the future lay in CNC machine tools, and particularly in the all-purpose machining centres, had sunk in and most firms were updating and redesigning their product line with this in mind. But time was important: they already lagged behind competitors and it took up to two years to design and develop a new range of products. When, therefore, the Japanese, fearing that voluntary export restraint would be extended, sought deals with British firms, they found a number of ready partners, and active encouragement from the British Government. The first deal in early 1983 was in fact between an American subsidiary (De Vlieg) and Okuma to make small machining centres. In mid-1983, another agreement was struck, between Bridgeport Textron (also American-owned) and Yasuda; this was followed later in 1983 by a three-year agreement between Frederick Pollard and Mori Seika. In 1984, Kearney and Trecker Martin completed a deal with Mitsubishi, and TI Matrix Churchill announced at the same time that they were going into partnership with Takisawa. Meanwhile, Yamazaki, one of Japan's largest privately owned machine tool manufacturers, announced investment plans for a £30m ($40m) automated plant in Britain, with the British Government making a £5m ($6.7m) grant towards its cost.[19] The philosophy was 'if you cannot beat them, join them'.

FRANCE

The indigenous machine tool industry in France has always lived under the shadow of the excellence of its German neighbour, although the French Government has long been determined to make it into a strong, independent presence within the French economy. Indeed, machine tools ranked as a high priority sector in the early French plans, with preferential allocations of steel and skilled manpower. The other major influence has been the growth of the automobile industry, with Renault and Peugeot both developing their own machine tool sections which have built up an independent presence in their own right.

By 1980, France had in fact created a notable export trade in machine tools, ranking eighth in terms of industrialised nations. Much of this surplus was, however, based on highly subsidised credit with Eastern Bloc countries and France was in fact a net importer of machine tools, with West Germany, Italy and latterly Japan as the main sources of supply (Rendeiro, 1984).

The Socialist Government came to power in 1981 to find the machine tool industry collapsing around it. The problems of the Polish economy led to the demise of the LINE group of companies, while other groups such as Huré (milling machines) and Ramo-Casaneuve (NC lathes) were also in trouble. The new Government's answer was a comprehensive machine tool plan which in effect created two large nationalised groupings (which meant, with Renault, that three nationalised concerns dominated the industry). Both involved extensive horizontal mergers. The MFL (Machines Françaises Lourdes) group became France's specialist producer of heavy-duty machine tools, concentrating on the whole on the traditional, one-off heavy-duty machine tools. It included much of the bankrupt LINE group, plus a number of other companies in the same field.

The other major 'pôle de développement' for the industry was Interlautomatisme encompassing Huré (specialised milling machinery), Graffenstraden (horizontal machining and flexible manufacturing systems) and H. Ernault Somua (HES) (lathes), the idea being to build this group up to provide the main French competitor with the Japanese in machining centres and NC lathes, and for it to move into flexible manufacturing systems. In fact it has never satisfactorily got off the ground, mainly because of the financial problems of the HES group. It is now likely that the group will be taken over by Toyoda, the Japanese machine tools group (part of Toyota with whom Renault already have a robotics agreement).[20]

Government involvement with the machine tool industry also meant heavy financial commitments. Both MFL and Intelautomatisme are now part-owned by Goverment through the IDI (Industrial Development Institute) and part by other French groups. The telecommunications group CIT—Alcatel, for example, has a 19 per cent stake in Intelautomatisme. The Government, however, have had to underwrite their losses. To date, neither MFL nor Intelautomatisme have made profits, although both predict profitability for 1986. Earlier estimates suggested that MFL was losing 35m frs per month ($5.3m) (Rendeiro, 1984). The main financial assistance for the industry has come from two soft loan facilities of 2.3 and 2.5bn frs ($350/380m), respectively available, first, for new investments by the machine tool industry itself and, second, for new investment in advanced computer controlled machinery. Access to these funds has been limited to firms which concluded planning agreements setting the usual targets for R&D, exports, training, etc. An additional soft loan facility of 1.2bn frs ($182m) was made available for the procurement of equipment for training schools.[21]

An analysis of EEC foreign trade in advanced machine tools

Competitive performance in international trade reflects—and has also given rise to—many of the technical and economic developments discussed above. There follows in this section a summary analysis of some elements in the evolution from 1976 to 1983 of the foreign trade of the leading EEC exporters of advanced machine tools.

The analysis is derived from Eurostat's foreign trade statistics for member countries which separated twenty-three categories of NC machine tools.[22] Quantities are shown in tons weight and values in ECUs (the European Currency Unit); the ECU has some advantages

Advanced machine tools 69

as a numéraire for international trade comparisons over the more commonly used US dollar, being less prone to the disturbing influence of the variations in individual exchange rates.[23]

In 1983, total EEC exports of NC machine tools, in tons weight, were hardly greater than in 1976—a rise of only 6 per cent (see Table 3.11). Meanwhile, the volume of world trade in all manufactured goods rose by 27 per cent. Over the same period, EEC imports rose in tons weight by as much as 67 per cent. The EEC export surplus was almost halved over the seven years, falling from 55 to 29 per cent of exports.

Several factors lie behind these developments.

Table 3.11 NC machine tools: EEC exports, 1976 and 1983

	Quantity '000 tons		Value Million ECUs	
	1976	1983	1976	1983
Belgium*	3.3	3.8	25.8	39.0
West Germany	23.4	37.5	211.8	636.9
France	2.3	3.8	28.7	60.6
Italy	7.5	12.6	59.3	153.7
United Kingdom	21.1	3.4	97.5	60.2
Total EEC (10)†	59.4	62.8	435.1	972.3

* Belgium-Luxembourg.
† Including other member countries. Excluding Greece in 1976.
Source: Eurostat, *Analytical Table of Foreign Trade*, 1976 and 1983.

EEC EXPORT PERFORMANCE

The first point to be observed is that the stagnation of total EEC exports of NC machine tools was by no means general. The main competitive developments *within* the EEC were the increasing dominance of West Germany, the substantial expansion in Italy and the striking decline of the United Kingdom's export business. West Germany's exports, indeed, increased in quantity by 60 per cent of the EEC total, from 39 per cent in 1976. Italy's exports also expanded, rising from 13 to 20 per cent of the EEC total. France and Belgium-Luxembourg increased their exports too. The total exports of these four countries rose in quantity by 58 per cent—an impressive performance in relation to the 27 per cent expansion of world trade in manufactures.

The collapse of British exports was dramatic, accounting by itself for the near stagnation of total EEC exports. In 1976, British exports were 36 per cent of the EEC total; by 1983 they were only 5 per cent. From being the second largest EEC exporter, the United Kingdom sank to the smallest among the five leaders (with lower exports than Belgium-Luxembourg). This melancholy performance can be contrasted with the leading position of the United Kingdom among West European countries in the early development of NC machine tools.

EXPORT PERFORMANCE AND 'TECHNICAL QUALITY'

The proposition is often advanced that, in these days of world-wide industrialisation, the longer-established industrial countries should safeguard their competitive positions by moving 'up-market': that is, by concentrating on products embodying the highest added value content and the most sophisticated technology.

Technical sophistication is difficult to measure objectively. One frequently used, although admittedly rough, statistical indicator of relative 'technical quality', applicable to comparisons of a particular product, is found in the recorded unit value of exports. For the present study, the value in ECUs per ton of exports has been calculated from the Eurostat data for NC machine tools. The question asked is whether recent experience in the EEC displays any association—positive or negative—between relative export performance and relative unit values in 1976 to 1983. Table 3.12 sets out the results.

The most striking change is the complete reversal of the British position, from the lowest unit value for the whole product group in 1976 (37 per cent below the average for the EEC as a whole), to the highest position in 1983 (14 per cent above the EEC average). The United Kingdom unit value nearly quadrupled, against an approximate doubling of the average for the EEC as a whole. There were less drastic changes in the rank order of the other exporters. West Germany remained second in rank and by 1983 the United Kingdom's place at the bottom was taken by Belgium-Luxembourg. France sank from top place to third place.

How far are the national differences in the changes in unit values associated with differences in export performance? As a first test of performance, we took changes in the main exporters' shares of total EEC exports (since the total increased, in quantity, by only 6 per cent, national changes in shares are not very different from changes in actual quantities). At this aggregate level, no unambiguous association

Table 3.12 NC machine tools: unit value of exports compared with shares of total EEC exports, 1976 and 1983

	Unit values of exports			Percentage of total EEC exports			
	'000 ECU per ton			In quantity		In value	
	1976	1983	1983 as %1976	1976	1983	1976	1983
Belgium*	7.9	10.2	129.7	5.5	6.1	5.9	4.0
West Germany	9.0	17.0	187.6	39.4	59.8	48.7	65.5
France	12.5	15.8	126.8	3.9	6.1	6.6	6.2
Italy	7.9	12.2	154.8	12.7	20.2	13.6	15.8
United Kingdom	4.6	17.7	383.5	35.6	5.4	22.4	6.2
Total EEC†	7.3	15.5	211.3	100	100	100	100

Note: In this and other subsequent tables, unit values, percentages, etc. are calculated from data reported with additional digits.
* Belgium-Luxembourg.
† Including other member countries. Excluding Greece in 1976.
Source: Calculated from source of Table 3.11.

appears (see again Table 3.12). The exceptional increase in the British unit value accompanied the extreme fall in the United Kingdom's share; by contrast, West Germany, with the second greatest increase in unit values, enjoyed the greatest increase in share of EEC exports. In France and Italy, moderate increases in unit values go with varying increases in shares. In Belgium-Luxemburg, a rather small rise in unit values accompanies a very small rise in share. For the exporters other than Britain, the association between moves up market (as represented by rises in relative unit values), and improvements in export performance (represented by increases in share of EEC exports), may appear positive, but cannot be regarded as a systematic association.

Broadly speaking, two types of explanation may underlie these differing changes. One possible explanation looks to domestic factors and the movement in exchange rates. The second takes into account the detailed changes in the composition, by kind of product, of national exports.

The first type of explanation might lie in the very different rates of domestic inflation of costs and prices, partly, but by no means wholly, reflected in fluctuations of exchange rates. To give some indication of such factors, Table 3.13 compares the increases in unit values expressed in ECUs with the same increases expressed in national currencies. For the United Kingdom, the very big rise in ECU unit values was almost matched by the rise in unit values expressed in sterling, suggesting that the main factor was the relatively fast rate of domestic inflation in 1976-83 (not reflected in the sterling value of the ECU, which rose by 6 per cent).[24] But the increase of 262 per cent in the unit values of NC machine tools, in sterling, was far greater than the doubling of producer prices for

Table 3.13 NC machine tools: unit values of exports in ECU and national currencies; 1983 as % 1976

	Unit value in ECU	ECU per unit of national currency	Unit value in national currency
Belgium*	129.7	0.95	136.5
West Germany	187.6	1.24	151.3
France	126.8	0.79	160.6
Italy	154.8	0.69	224.5
United Kingdom	383.5	1.06	362.0

*Belgium-Luxembourg.
Source: Calculated from source of Table 3.11.

manufactures as a whole; general domestic cost and price increases thus explain only part of the big rise in unit values of British NC machine tools. In West Germany, by contrast, the rise in unit values in DM of 51 per cent was little more than half the 90 per cent rise in ECUs (although rather more than the 30 per cent rise in producer prices of all manufactures), indicating, again, that other factors were at work.

The second type of explanation of changing unit values looks to changes in the composition of exports in terms of 'quality'. How far have aggregate unit values concealed important changes in the types of machine tools exported, which may indicate changes in 'technical quality'?

For a summary measure of changes in 'technical quality', each country's exports of the twenty-three categories of NC machine tools distinguished in the Eurostat trade figures have been condensed into three groups, based on the rank order of unit values of *total* EEC exports of the twenty-three categories in 1983. The three groups are labelled 'up-market', 'medium', and 'down-market' (noting, however, that all NC machine tools could be regarded as 'up-market' by comparison with machines not so equipped). In the up-market group of NC machine tools, the average unit value is 20 per cent above the average for total EEC exports, and all the categories included are at least 12 per cent above this average. The medium group averages 4 per cent below the overall average. The down-market categories average 35 per cent below the overall average, and all are at least 25 per cent below.[25] The results are shown in Table 3.14.

Some results stand out from this exercise:

(a) The high unit value of the United Kingdom's total exports in 1983 reflects the much higher proportion (55 per cent) of exports in the up-market categories than the EEC average (44 per cent); and the big rise in unit values is reflected by the remarkable increase in the proportion of exports in this up-market group (only 26 per cent in 1976). But export performance in this up-market group was only very slightly less depressed than in the other categories; the British share of EEC exports in the up-market categories fell from 24 to 7 per cent—moderate only by comparison with the fall from a dominating 58 per cent of the down-market sector in 1976 to only 5 per cent in 1983. In short, the United Kingdom's loss of its very strong position down-market was only slightly offset by retaining a slender toe-hold in the up-market categories.

Table 3.14 NC machine tools. Composition of exports compared with shares of total EEC exports (both in tons) and export unit values

	Market group*	Percentage composition of exports		Percentage share of total EEC exports		Unit values of exports '000 ECUs per ton	
		1976	1983	1976	1983	1976	1983
Belgium†	U	26.5	58.0	3.7	7.9	7.9	12.2
	M	27.3	26.8	4.6	4.6	8.2	9.2
	D	46.2	15.2	9.1	4.6	7.7	4.4
	Total	100	100	5.5	6.1	7.9	10.2
West Germany	U	58.0	44.0	57.7	60.1	9.3	20.3
	M	30.6	34.5	37.0	58.4	8.8	16.1
	D	11.4	20.7	16.2	61.4	8.1	11.2
	Total	100	100	39.4	59.8	9.0	17.0
France	U	34.9	44.6	3.4	6.1	15.7	16.2
	M	32.3	39.5	3.8	6.8	14.7	18.3
	D	32.9	16.0	4.6	4.8	6.8	8.5
	Total	100	100	3.9	6.1	12.5	15.8
Italy	U	30.0	36.7	9.6	16.6	8.3	16.1
	M	54.9	43.9	21.4	25.0	8.6	11.5
	D	15.1	19.4	6.9	19.3	4.2	6.2
	Total	100	100	12.7	20.1	7.9	12.2
United Kingdom	U	26.4	55.5	23.7	6.8	6.2	20.0
	M	27.8	25.1	30.4	3.9	5.5	15.7
	D	45.8	19.4	58.5	5.2	3.2	13.7
	Total	100	100	35.6	5.4	4.6	17.7
Total EEC‡	U	39.6	44.5	100	100	8.7	18.5
	M	32.6	35.3	100	100	8.0	14.7
	D	27.8	20.2	100	100	4.6	10.1
	Total	100	100	100	100	7.3	15.5

* U = 'up-market' categories; M = 'medium' categories; D = 'down-market' categories. These categories are derived by grouping the 23 categories in the source according to unit values of total EEC exports in 1983 (see text).
† Belgium.
‡ Including other member countries. Excluding Greece in 1976.
Source: Calculated from source of table 3.11.

(b) The *West German* experience was very different. The share in the up-market sector actually declined (from 58 to 45 per cent of EEC exports), while West Germany's share in the down-market group increased strongly (from 16 to 61 per cent). In the down-market sector, West Germany replaced Britain as the dominant exporter and West Germany's overall improvement in export performance was almost wholly due to the capture of this much increased share of the down-market categories. Yet in practically every category, up-market or down-market, West Germany's unit values were above the EEC average.

(c) The *French* experience was different again; the rise in the French share of total EEC exports was almost entirely the result of increased shares in the up-market and medium categories.

(d) The strengthened position of *Italy*—from 13 to 20 per cent of total EEC exports—was due to increasing shares of *both* up-market *and* down-market categories but, as in West Germany, was most conspicuous in the down-market sector.[26]

The association (or lack of it) between relative unit values and export performance in 1983 can be analysed in another way. For each exporting country we calculated the rank correlation between:

(a) The rank order of the country's unit values by category, as ratios to the total EEC unit value for that category (indicating the extent to which the country's exports in each category are more, or less, 'up-market' than the EEC average for that category).

(b) The rank order of each country's share of EEC exports for each category, in value, as a ratio of that country's overall average share (this is one commonly used measure for displaying the items within a group of products for which the country holds a 'revealed comparative advantage').

The rank correlations between these two variables emerged as follows:[27]

Belgium-Luxembourg	+0.04 (not significant)
West Germany	−0.85 (significant at 1%)
France	+0.51 (significant at about 6%)
Italy	+0.70 (significant at 1%)
United Kingdom	+0.04 (not significant)

These results again display the variability of the relationships between 'technical quality' and export performance. The negative correlation for West Germany confirms the showing of the earlier analysis, that West Germany's competitiveness is most marked in the down-market categories. Italian and French experience is more mixed, with some tendency, especially in Italy, for comparative advantages to be revealed in the more up-market categories. For Belgium-Luxemburg and Britain no relationship emerges.

From these statistical exercises, the conclusion must be that concentration at the upper end of the market for NC machine tools may have helped Britain to preserve a small niche in the market, and France—and perhaps Belgium-Luxembourg—to make some competitive gains. But West Germany has, in the last few years, benefited more by expansion in lower-valued categories; and Italy has gained market shares at both ends of the market. Shifts upmarket are not, it may be suggested, the royal road to competitive success in exports of NC machine tools. All depends (to state a platitude) on the capacity of the industry to adapt technical qualities, prices and marketing to market needs. The West German experience, in particular, seems to illustrate the opportunities still open, despite an increasingly competitive market, for creating a strong position in the relatively down-market categories of NC machine tools. How long this strong position can be held is the open question.

IMPORT COMPETITION IN THE EEC MARKET FOR NC MACHINE TOOLS

This analysis has concentrated on export performance, with only a glancing reference to the 67 per cent increase in the quantity of the EEC's imports between 1976 and 1983. This increase was mainly in imports from outside the EEC (which rose in quantity by 123 per cent), while intra-EEC trade increased by only 29 per cent (according to the import statistics).

The massive appearance of Japan on the EEC market for NC machine tools was the most important element. From only 12 per cent of total extra-EEC imports in 1976, imports from Japan rose to 46 per cent by 1983. Other significant import suppliers, Austria, Sweden and Switzerland, achieved increases in quality, but the United States lost ground. The nature of the pressure of imports, especially from Japan, can be illustrated, again, by analysis of unit values (see Table 3.15). The main feature is a 13 per cent lower average unit value of imports from Japan than that of EEC total exports in 1983; in fact, since imports are valued CIF against FOB export values, the FOB/FOB gap might be put at nearer 20 per cent. There was no such gap in 1976. Over the seven years, the unit value of imports from Japan rose by only 50 per cent, compared with the doubling of EEC export unit values.

How far is this apparent incursion of down-market competition from Japan confirmed by the composition of the Japanese products? It may seem paradoxical that the bulk of imports from Japan is *not* found in the products which we have defined, on the basis of EEC exports, as down-market. Three-quarters, in quantity, of the imports from Japan in 1983, fall in our 'up-market' group and only 7 per cent in the down-market sector. The clue appears to be that the low unit value of the Japanese products is most marked in the up-market categories. Within the up-market categories, the biggest price gap lies in one important type alone: NC jig-boring machines, which accounted for 31 per cent of the quantity of imports from Japan in

Table 3.15 NC machine tools: EEC imports from non-EEC countries, 1983

	Market group*	Quantity '000 tons	Unit value (CIF) '000 ECUs per ton
Total from non-EEC			
	U	14.1	16.6
	M	7.2	14.0
	D	3.0	13.7
	Total	24.3	15.5
of which—			
from Japan			
	U	8.4	13.9
	M	2.0	12.9
	D	0.8	9.6
	Total	11.2	13.4
from Switzerland			
	U	2.2	32.5
	M	1.6	26.6
	D	0.8	16.6
	Total	4.6	27.6

* U = 'up-market' categories; M = 'medium' categories; D = 'down-market' categories. These categories are derived by grouping the 23 categories in the source according to unit values of total EEC exports in 1983 (see text).
Source: Calculated from source of Table 3.11.

1983, with a unit value (CIF/FOB) as much as 38 per cent less than that of EEC exports of this type. It may be conjectured that this high concentration of low-priced Japanese machines represents the partial replacement of a fairly high-priced category in the EEC's export range. The other big item in imports from Japan (39 per cent of the total quantity) is another key product, NC lathes, although in this case the Japanese unit value is only 7 per cent less than that of EEC exports. In both these categories, the Japanese development of low-cost standardised machines seems to be the main factor in their competitive success.

Apart from Japan, the most significant supplier of imports from outside the EEC is Switzerland, with 19 per cent of extra-EEC imports (the quantity having more than doubled since 1976). As could be expected, the imports consist mainly of up-market and medium machines with unit values approaching twice those of the corresponding EEC exports; demand for the specialised high-price Swiss products has been well maintained.

REACTING TO JAPAN'S COMPETITION

These analyses illustrate in quantitative terms many of the earlier observations in this chapter stressing the differing strategic reactions of national industries to the rising tide of competition. West Germany is the largest EEC importer of NC machine tools, accounting for over one-third of EEC imports from Japan (Table 3.16). But the West German industry, being also the largest producer and exporter, has

Table 3.16 NC machine tools: imports by principal EEC exporting countries, 1983 ('000 tons)

	Total from all sources*	From Japan
Belgium†	5.71	2.90
West Germany	13.36	4.11
France	8.73	0.73
Italy	3.78	0.40
United Kingdom	8.01	2.06
EEC total‡	44.86	11.14

* Including imports from other EEC countries.
† Belgium-Luxemburg.
‡ Including other member countries.
Source: As for Table 3.11.

so far had the strength to meet Japanese competition by adjustment of the product range with relatively little disruption of the growing market. The Italians to some extent followed the same path. The market losses of the British industry, on the other hand, demonstrate failure to exploit, by adjustments of design and production, the earlier British advantages in the lower-priced categories. As already observed, there are now indications that the lessons have been learned and that British producers, in part by working with the Japanese, are preparing modernised product lines which should, in the course of time, offer opportunities for improved trade performance.

Conclusions

Taken as a whole, the machine tool industry does not really fit the criteria for high technology industries; R&D intensity is low and growth rates sluggish. Technological change is concentrated in those firms and sub-sectors which make intensive use of advanced microelectronics and experiment with new industrial materials. *But these sub-sectors are becoming increasingly important to the industry.* Hence the interest of this chapter. Electronics initially affected only the periphery of the industry—the highly specialised, custom-built machine tools for the aerospace market. As time has gone by, this 'new industrial activity' has gradually made deeper and deeper incursions into the main body of the industry until, today, for the European machine tool industry, CNC technology has become not just an addition to the top end of the product range but an essential lifeline for survival.

The most important structural trends caused by the emergence of electronically controlled machinery in producing and using industries can be summarised as follows:

— The importance of scale economies in the production of machine tools is increasing as a result of the advance of general purpose machines.
— Customer-orientated design, which demands intimate knowledge of user needs, is of declining importance.
— While the cost of electronic controls is decreasing, the increasing precision which electronics can bring demands industrial materials of higher quality.
— Scale economies in the use of electronically controlled machine tools are decreasing as the tools themselves become more flexible.
— Manpower requirements are changing. In the past, the availability of metalworking skills was essential. Nowadays it is knowledge of the application of electronics that has become the prerequisite for success.

The implications of these trends for Europe are far-reaching. European producers formerly had considerable advantages in terms of craftmanship in the labour force and the intimate knowledge of customers' needs. Neither of these advantages counts for so much today. The increasing market penetration of the general purpose machine tools means that the market is becoming more price competitive and new firms are entering the industry from adjacent sectors—for example, electronics. In consequence, there is increasing internationalisation of production which has in turn brought increasing government involvement in the industry.

It is essentially the challenge of Japan, not that of the United States, that now concerns Western Europe. The Japanese challenge has a new dimension, because Japanese developments in CNC machine tools and industrial robots were commercially orientated from the very beginning. In contrast to the performance of American producers, Japanese firms have gained considerable ground in world markets. They have done so by concentrating on a limited range of machine types. Their competitiveness in this limited range is well illustrated by the unit value analyses of the previous section.

In Western Europe, and particularly West Germany—the dominant producer—it appears that firms failed at first to recognise the opportunities offered by the application of microelectronics to machine technology. They were relatively efficient in mechanical precision engineering, and seem to have been reluctant to take up new electronic devices for this very reason. And what was at one time an advantage—attention to detail and willingness to fit in with customer requirements—proved to be a major barrier to rapid technological adaptation.

In recent years, European firms have been catching up. In West Germany, for example, the rapidly rising share of CNC machine tools in total NC machine tool production between 1977 and 1982 is testimony to the pace of change, while the early 1980s have witnessed the take-off phase in robotics with estimates of 7,500 robots installed by the end of 1985, compared with 1,200 in 1980. The West German machine tool industry has shown a remarkable ability to respond to competitive pressures. It is interesting, for example, that the detailed analysis of the trade statistics supports the qualitative picture of the industry meeting the Japanese competition head-on by moving down-market into series production of CNC lathes and machining centres. Yet the success story of Europe is perhaps Italy rather than West Germany, for it was Italy which, at an early stage in CNC development, recognised the market potential for the general-purpose computer controlled machine tool, admittedly behind a wall of protection. Helped by the major links forged with the United States, the Italians are now in a better position than any of their EEC counterparts to move into full-scale automation.

The disaster story of Europe has been the British machine tool industry. In the 1960s, the British were well ahead of their European competitors in both knowledge and use of NC technology. The virtual collapse of this advantage, shown so dramatically by the fall in their share of EEC NC machine tool exports from 36 per cent to 5 per cent in the seven years 1976 to 1983, is sad testimony to the deep-rooted ills of the British economy. As in the case of computer aided design, it illustrates the British ability to grasp highly technical, highly sophisticated new technologies, but their abject failure to diffuse them more widely—a failure both of marketing (indeed even to appreciate the market opportunities that the technology offered, which both the Italians and the Japanese understood so well) and of technological competence, with the lack of skilled technicians seemingly the factor most severely inhibiting the diffusion of the technology in the United Kingdom. Government policies in the United Kingdom, aimed primarily at the diffusion of NC technology, have in fact brought the NC share of the total machine tool park in Britain well up to the level of its competitors, but it appears that lack of skilled technicians inhibits full advantage being taken of this position (Daly, Hitchens and Wagner, 1985). Moreover, these measures give no advantage to British-produced machine tools. The detailed trade statistics indicate that the British producers have now retreated to a highly sophisticated niche in the NC market: it now appears that they are hastening to fill the gaps in their product range by joint ventures with the Japanese; but such a lifeline does not necessarily help them cope with the next phase of this revolution, namely full-scale automation.

Assessing the effects of government intervention is difficult, for various reasons. First, the existence of technical externalities which could give government a proper role in the game, are never easily identified. Second, the funds for government aid must be raised by taxes; these impose costs (disincentives) on the community which may be widely spread but are real. Third, the presence of government intervention (e.g. R&D subsidies or trade protection) changes the incentive structure and competitive conduct of favoured firms. It is impossible to know what the effects would have been in the absence of government interference. Thus, even in the case of apparent successes of policy, the ground for judgement is shaky.

With regard to machine tools, the most prominent role of government has been to accelerate technological developments to meet military demand for precision engineering, above all in the aircraft industry. The 'visible' hand played a role in the case of the first NC machine tools as well as for CNC machine tools and industrial robots. But military R&D does not necessarily mould into lasting commercial success. The United States, the United Kingdom and France—the 'big league' in military research—are not the countries with a strong competitive edge in advanced machine tools.

Selective government intervention in machine tools *per se* has not been particularly intensive in most industrial countries. Britain (in the 1960s) and France (in the 1980s) have both tried to 'rationalise' their industries, putting emphasis on larger size and a limited, complementary product range. The British experiment gradually disintegrated as it became apparent that flexibility, not size, was the competitive requirement for the 1970s. It is too early to judge how successful the French attempts at rationalisation will be. Already we have noted how technological developments have increased the importance of size in the production of series CNC machines. It is arguable that the shift towards full-scale system automation will accelerate this trend, demanding resources for design and development well beyond the capacity of the small or medium-sized firm typically found in the machine tool industry.

A salient feature of CNC machine tools and industrial robotics is the extent to which these industries rely upon the electronic industry for their high technology content. For this reason, the involvement of governments in this high technology area, especially in semiconductors, is relevant to the advanced machine tool sector. Indirect spin-offs from government promotion of micro electronics, and from military and space programmes, have frequently occurred. The move of electronic firms into the production of CNC machine tools and, particularly, robots, to be observed in Europe as well as in Japan and the United States, has reinforced these links. To look only at direct government involvement in CNC machine tools and industrial robots is therefore to underestimate true involvement.

It has to be borne in mind that in the machine tool industry it has always been relatively easy to obtain licences from abroad, and a large number of such agreements have been made. By itself, this would appear to diminish the role of government. The rise of the Japanese machine tool industry is a case in point; licences from Western Europe and, later on, from the United States, played an important role in the development of the industry.

The issue of international licensing has to be seen against the fact that until some years ago direct international investment and international co-operation among firms were of minor importance. In this respect things have changed considerably. The internationalisation of machine tool firms via foreign investment and co-operation agreements has proceeded fast. Large firms have been particularly active in this field. Two main factors seem to have triggered this recent development. First, there is increasing fear on the part of Japanese firms (backed by past experience) that import restrictions could be imposed in the future. Second, there has been a trend toward standardisation (towards so-called general purpose 'catalogue-machines') which has eliminated the old links between customer and manufacturer. Design, assembly, component manufacture and marketing activities can increasingly be split, thus widening the scope for international relocation of production and international subcontracting.

If these trends towards internationalisation hold, it is difficult to predict what place the Japanese will occupy. There are some question marks over their current position. First, it remains to be seen how far the trend towards general purpose machinery (where the Japanese strength lies) will continue or whether it will be halted by full-scale automation which requires an individual package of hardware and software tailored to each customer's need. Second, electronic control devices have become so cheap and readily available that anyone knowlegeable in the area can enter: but the opportunities for application must be sought out and European firms have not always shown themselves adept in this search. Third, to be successful, the scope for precision engineering offered by electronic control often requires new industrial materials for mechanical parts of machines. European firms have traditionally had a technological advantage in this area; but whether they can retain it is open to question.

Perhaps the most abiding conclusion to be derived from this chapter is the durability and flexibility of the industry itself. Confronted by the trauma of deep recession and a major competitive challenge from Japan (and in traditional machine tools from the NICs), the industry has survived (even in the United Kingdom it is surprising how many firms survived, albeit in much reduced form) and, in West Germany and Italy, has responded by establishing new areas of strength. The two countries which have been least successful in this respect have been France and Britain—both countries in which governments have played, or sought to play, a significant role in the adjustment process. It is always difficult in such circumstances to identify cause and effect. West German experience seems to emphasise the inherent capacity of firms to adjust to market pressures, whereas proponents of government intervention may argue that well-designed intervention could be effective if it were not so often too little and too late. But governments frequently, and sometimes with good reason, climb aboard only when the train is already moving. Since a government has limited capacity to restart a stalled locomotive, it may well be that such measures have better chances of success when they coincide with the beginning of a spontaneous adjustment process.

Notes

1. Computer numerical control means that a control unit is freely programmable, and attached to the respective machine tool. Numerical control does not offer this flexibility.
2. Here and in the rest of the chapter, a narrow definition of robots is applied.

Robots are defined as handling or working machines freely programmable for different kinds of work, with two or more motion axes (OECD, 1983a), not merely automated manipulation.
3. Another argument, although not always stressed in official documents, rests on defence considerations: machine tools are an important equipment for weapons production and governments are reluctant to rely on imported machinery. The stated objectives very often refer to assertions like the one that a technologically progressive machine tools industry is essential for innovation in the whole manufacturing industry.
4. Of course, the figures on NC machine tools installed in different countries should be interpreted with caution. The definitions applied and the accuracy of the estimate can be expected to differ.
5. The usual Japanese definition is much broader. Therefore, official figures must be revised downward to make them comparable to European data. According to the Japan Industrial Robot Association, an industrial robot 'is a machine capable of performing versatile movements resembling those of the upper limbs of a human being or having sensory and recognition capability and being capable of controlling its own behaviour'. International comparisons based on national figures with diverging definitions therefore very often resemble comparisons of random data.
6. *The Financial Times*, 19 January 1982, 'West German Machine Tools', Kevin Done.
7. Horn, E.-J. (1982), *Industrial Adjustment and Policy: IV. Management of Industrial Change in Germany*, Sussex European Papers No. 13, University of Sussex.
8. National Academy of Engineering (1983), p. 33.
9. See latest figures quoted in *The Financial Times*, 19 February 1985: 'World machine output grows by 6.5 per cent'. Italy's share of world exports from these figures appears as 6.5 per cent, compared to the American share which had fallen to 4.5 per cent.
10. See *The Financial Times* supplement on Italian engineering, 27 September 1983, 'Factory Automation brings benefits to sector', p. iv.
11. See *The Financial Times* supplement on 'Manufacturing Automation', 5 February 1985, p. 22. Profile on Comau.
12. Britain's Machine Tool Industry: The Uphill Struggle to Survive', Ian Roger and Peter Bruce, *The Financial Times*, 15 February 1983.
13. Board of Trade Commitee under Sir Stuart Mitchell, Report on The Machine Tool Industry. HMSO, London, 1960. See also C. T. Saunders (1978), *Engineering in Britain, West Germany and France*, Sussex European Papers No. 3, Sussex University.
14. Prais, S. (1981) estimates that the total committed to the machine tool industry over the period 1966-78 was about £100m.
15. See note 12 above.
16. See 'Metalworking Production: Fifth Survey of Machine Tools and Production Equipment in Britain', July 1983. The figure given in the survey is 3.32 per cent, which puts Britain on a par with Japan and the United States. See also notes to Table 3.7.

17. See Daly, Hitchens and Wagner (1985).
18. See Nasbeth and Ray (1974), p. 31.
19. *The Financial Times*, 4 July 1984, supplement on British engineering, 'Steady Growth in New Orders', Peter Bruce and *The Financial Times* 27 July 1984, 'Yamazaki to invest £30m in UK plant', Peter Bruce.
20. *The Financial Times*, 10 January 1985, 'French machine tool makers see brighter start for 1985', David Marsh.
21. See Note 20.
22. The source is Eurostat, *Analytical Tables of Foreign Trade*, 1976 and 1983, based on the NIMEXE commodity classification which separates twenty-three categories of NC machine tools, quantities being given in tons and values in ECUs.
23. The exchange rate of the ECU was $1.2 in 1976 and $0.89 in 1983.
24. The 16 per cent fall of sterling against the US dollar over this period is less relevant. More relevant for British trade is the small fall, of 3 per cent, in the 'effective', or trade-weighted value of sterling.
25. The lack of symmetry in this grouping is imposed by the very uneven distribution of trade between the twenty-three categories. In addition, a number of groups were affected by the fact that in 1976 the United Kingdom, then a new entrant to the EEC, was unable to give separate data for some NIMEXE items.
26. The apparent up-market shift in the exports of Belgium-Luxembourg may be deceptive. It is due mainly to the increase in one up-market category— jig-boring machines—which accounted for over half of Belgium-Luxembourg exports in 1983 but were sold at unit values well below the EEC average for this category, and shared this market at low prices with the Japanese.
27. These rank correlations were calculated from the twelve categories in which EEC exports in 1983 exceeded 20m ECUs.

Bibliography

Batelle-Institut, Frankfurt (1982), *Soziale Implikationen der Einführung von Industrie-Robotern im Fertigungsbereich.* Unpublished.

Commission of the European Community (1983), *The European Machine Tool Industry*, SEC (83) 151, Brussels.

Daly, A., Hitchens, D. and Wagner, K. (1985), 'Productivity, Machinery and Skills in a Sample of British and German Manufacturing Plants', *National Institute Economic Review*, February, pp. 48–61.

Daly, A. and Jones, D. T. (1980), 'The Machine Tool Industry in Britain, Germany and the United States', *National Institute Economic Review*, No. 92, May, pp. 53–63.

Junne, Gerd, and van Tulder, Rob, (1984), *European Multinationals in the Robot Industry*. A Pilot Study. Universiteit van Amsterdam, mimeo.

Nasbeth, L. and Ray, G. F. (1974), *The Diffusion of New Industrial Processes*, Cambridge University Press.

National Academy of Engineering (1983), *The Competitive Status of the US Machine Tool Industry*, Office of the Foreign Secretary, US National Academy of Engineering.
OECD (1982), *Micro-electronics, Robotics and Jobs*, Paris.
OECD (1983a), *Industrial Robots. Their Role in Manufacturing Industry*, Paris.
OECD (1983b), 'Robots: The Users and the Makers', *The OECD Observer*, No. 123, July, pp. 11-17.
OECD (1984a), *Trade in High Technology Products. An Examination of Trade Related Issues in the Machine Tool Industry*. DSTI/SPR/83. 102, OSTI/IND/83. 40 (2nd rev.), Paris, 22 March.
OECD (1984b), *Trade in High Technology Products*. Background Report on the Method of Work and Findings of the Studies carried out by the Industry Committee and the Committee for Scientific and Technological Policy. DSTI/SPR/84.1, DSTI/IND/84.7, Paris, January.
Prais, S. J. (1981), *Productivity and Industrial Structure*. A Statistical Study of Manufacturing Industry in Britain, Germany and the US, Cambridge, Cambridge University Press.
Ray, George, F. (1984), *The Diffusion of Mature Technologies*, Cambridge, Cambridge University Press.
Rempp, Helmut (1982), 'The Economic and Social Effects of the Introduction of CNC Machine Tools and Flexible Manufacturing Systems', *European Employment and Technological Change*, European Centre for Work and Society, New Patterns in Employment, No. 2, Maastricht.
Rendeiro, Joao, O. (1984), 'Technical Change and Strategic Evolution in the Machine Tool Industry', *Sussex European Research Centre*, mimeo.
Rosenberg, Nathan (1963), 'Technological Change in the Machine Tool Industry', *The Journal of Economic History*, 23, pp. 414-46.
Saunders, Christopher, T. (1978), *Engineering in Britain, West Germany and France*, Sussex European Papers No. 3, Brighton, the University of Sussex.
Schenk, Winfried (1982), 'Industrieroboter—Eine Chance für Österreich?', paper presented at the international meeting on industrial robots. Linz, September 28-30.
Verband deutscher Maschinen und Anlagenbau (VDMA) (1983), *Der Industrieroboter*. Unpublished paper.
Wolfsteiner, Manfred (1983), 'Einfluss der Robotertechnik auf Beschäftigung und Tätigkeiten', *Mitteilungen aus der Arbeitsmarkt und Berufsforschung*, 16, pp. 167-76.

4 Telecommunications: a challenge to the old order

Godefroy Dang Nguyen

Telecommunications is not a new industrial activity. The telephone has now been with us for over a century and by the 1960s was to be found in almost half the households of Western Europe. The advent of information technology has, however, thrown the industry into new prominence. The microprocessor in essence enables the enormously rapid processing of huge amounts of information: the telecommunications system is the means by which all this information can be transmitted from one point to another. It therefore becomes a vital part of the infrastructure of any new 'high technology' society.

The microprocessor also extends the capability of the telephone and telephone system. Like so much else, it enables it to become 'intelligent'—and this in turn has added to the range of services which can be provided via the system. Hence the term 'telematics'. Yet, so far, surprisingly little of this has actually percolated down to the average European consumer. Ninety per cent of revenues are still derived from straightforward telephone calls, and in spite of all the publicity given to information technology, the typical domestic consumer remains unaffected: his telephone for the present is a means of voice communication only.

In this chapter we shall look behind this façade of apparently slow developments to investigate the changes that have been taking place and the impact they are having. The focus is Europe, and specifically the four major industrial nations of Europe—France, West Germany, Britain and Italy. These four countries have shared a common institutional framework in which the network or service facility has been the responsibility of a public monopoly frequently linked to the postal services (hence the name PTT which we shall use to denote these authorities), while the equipment used by the networks (for terminal and switching facilities) has been supplied by private enterprise (although, inevitably, given the need for standardisation throughout the network, the relationship between PTT and equipment suppliers has been very close). The size of the market in all these countries is such as to provide a large and secure home base for these manufacturers, and the PTT, through its procurement policies, inevitably finds itself in effect running an industrial policy.

The telecommunications network consists of three separate operations. Through cables, satellites or microwaves, the voice is transmitted from one terminal to another; but given the number of subscribers, nobody can keep a line permanently open, hence the necessity of a switching system, which opens up a circuit for a subscriber when he requires it. The three functions—transmission, switching and terminal communications—correspond to three types of equipment. Until recently, the main characteristic of this equipment has been its long life expectancy. Switching equipment has lasted 20 to 30 years and transmission equipment 10 to 20 years. New equipment was therefore only gradually taken up to replace the old, and this has resulted in great inertia in terms of technology change.

Writing in 1977, Cherry (1977) described the 100-year-old telecommunications industry as 'a milieu' which has to run 'the most complex machinery that has ever been built'—namely the telephone network. The industrial adjustment of the industry in the last decade or so is the story of the adaptation of this complex machinery to rapid technological change. To help understand it, it is worth looking briefly at the history of the industry.

The telephone was invented in the United States by Alexander Graham Bell[1] in 1876 and his company initially struggled to maintain its position alongside the telegraph monopoly of Western Union. Once established, however, the company grew fast, protected by Bell's patents. When these expired there was a period of fierce competition with as much as half the network being controlled by 12,000 small companies (Libois, 1983). The Bell company, AT&T (American Telephone and Telegraph Corporation) fought back through the steady acquisition of its small competitors. An anti-trust case in 1913 guaranteed them interconnectability, but AT&T continued to grow and by 1932 controlled 79 per cent of the network (Brock 1981, p. 157). AT&T had its own company for manufacturing equipment, Western Electric, whose strength came not only from the huge demand of the AT&T network itself but from a strong innovation policy, with many new products emerging from the renowned Bell Laboratories. A second anti-trust suit in the 1920s again sought to bridle the power of AT&T, and it was forced to sell off its overseas subsidiaries (most of which subsequently formed ITT—International Telephone and Telegraph Corporation) and Western Electric was restricted to sales within the United States. An agreement was reached with ITT giving the latter the exclusive right to AT&T patents overseas. ITT also acquired the laboratories of International Western Electric in Britain, France and Belgium.

The development of the telephone system in Europe was very different. In the United Kingdom, Bell established a subsidiary in 1878 which, as in the United States, had to compete with the telegraph. But the telegraph system in the United Kingdom was run by the Post Office (PO). At first the company was forced to pay a royalty to the PO, the latter maintaining its monopoly over long-distance transmission, but in 1891 the PO itself entered the telephone market and eventually took over Bell and the other private companies in 1911.[2]

In France, the strength of Napoleon's semaphore system, which had established a nation-wide system of communication as early as 1800, made it difficult for both telegraph and telephone to get themselves established, and private companies struggling to establish both systems were subject to stringent regulations and eventually nationalised—again as part of the postal service—in 1889. In Germany, the Reichspost did not wait for private competition: the public monopoly on the telegraph was extended to the telephone system in 1881. In both Sweden and Italy the private sector lasted somewhat longer. In Sweden the whole system was not nationalised until 1918: in Italy, parts of the system remain private even today.

By the 1920s, therefore, the pattern was set. In the United States, there was a publicly regulated private monopoly with its own manufacturing subsidiary supplying equipment: in Europe there was public monopoly control of the network, with equipment supplied by private enterprise firms. It had taken roughly thirty years for the industry to stabilise but this stabilisation proved durable and lasted for fifty years.

What is happening to the industry today has its analogies with the past: once again innovation is encouraging new entry and challenging the position of a monopoly. This time the rivalry is between the telephone system and developments in telematics rather than between telephone and telegraph, but once again the issue of interconnectability between competing networks is a major one in the preservation of the monopoly.

The background to present developments

From the 1920s onwards the telephone systems grew steadily in both North America and Europe, although in Europe only Sweden and Switzerland could keep pace with American growth rates and

Table 4.1 Telephone penetration rates by country

	Connections per 100 inhabitants	
	1967	1982
Sweden	37	60.2
Switzerland	25.9	46.5
Canada	27.2	42
United States	30.5	41
FRG	9.2	37.3
Netherlands	12.9	36.9
France	6.6	35.6
Japan	11.7	35.4
United Kingdom	11.6	34.5
Belgium	11.8	27.9
Italy	9.1	25.9

Sources: Rugès (1970); Hoare and Govett (1984).

penetration was particularly slow in France (see Table 4.1). On the equipment side came the introduction of automatic switching with the Strowger process competing with the AT&T Rotary. The British PO opted for Strowger, and the Deutsche Bundespost (DBP) chose a version of the Strowger process developed by Siemens and Halske. But ITT, marketing Rotary, won orders in France, Belgium and Spain. These technological choices influenced the relationship between the network operators (the PTTs) and the manufacturers for fifty years, in the sense that, given the long life of this equipment, the network operators were tied into these manufacturers' systems for a long period.

After the Second World War came a wave of new developments—automatic transmission (the Crossbar system), multiplexing, radio transmission and then, in the 1960s, the first attempts at electronic switching and satellites. But these developments left the established structure of the industry unchanged. This was because the system, as we have seen, combined the complexity of the network with the provision of a single, relatively simple, service—namely voice communication. This latter characteristic meant that the market could not easily be fragmented between different types of equipment use (as is happening now with the mushrooming of telematics services) and in this sense was a classical case of natural monopoly. As far as network operators were concerned (in Europe, the public PTTs), priority lay with the unity of the network: new equipment had to be compatible with old and meet system standardisation requirements. This in itself gave the 'incumbent' supplier a market advantage, but it was further reinforced by the PTTs who generally responded to the need for compatibility by limiting their orders to a small group of firms with whom they established close links.

Among manufacturers, three in particular have taken advantage of this situation: ITT, Ericsson and Siemens. When a country decided to develop or modernise its network (for example with automatic switching), these three were always on hand with offers of help with development, and often established subsidiaries (or bought up local manufacturers) in order to appear to be home producers. Once the first contract was won, the need for maintenance, if nothing else, effectively guaranteed a continuing presence for the company in that market. The present market situation is the consequence of this competitive process, which, starting in the 1920s, has stabilised market shares for forty years.

A distinction is usually made between three categories of market (Jequier, 1976; Little, 1983):

(i) *The protected markets* where there are often several national manufacturers, sometimes including the subsidiary of a multinational company, with privileged access to the PTT's orders. In that group one finds, among others, the United States, Canada, Japan, France, Germany, the United Kingdom, Italy, Belgium, Sweden and Spain. In 1976 these countries accounted for more than 90 per cent of world production. Their firms are characterised by the fact that they have mastered the development of the technology and do not need to look for external transfer of technology.

(ii) *The open market*, which consists of countries which have no manufacturing capacity, most of them being in the Third World. It currently represents only 11 per cent of world markets but it is a fast-growing segment, with growth rates of 8–10 per cent against the 6–8 per cent world average. By the end of 1990, it will have a 12 per cent share of the world market. The most significant market for the next ten years will be Latin America, with 58 per cent of the open market, followed by Asia, with 29 per cent (Sarati, 1984). Most of these countries aim to establish their own manufacturing capacity but they lack the technological capability and therefore have to look to one of the major manufacturers. The latter obtains orders for the local PTT in exchange for installing a plant in the country. Competition for such contracts is particularly strong among manufacturers.

(iii) *Socialist countries*: in many respects, this market is similar to the previous one. It has an expected growth rate of 14 per cent a year and its present size is around $9 billion (15 per cent of the world market). The socialist countries lag far behind the Western countries, as far as telephone density is concerned. They buy Western technology but (as much as countries from the open market) want to have their own manufacturing units built with the help of large manufacturers.

Details of current regional markets for telecommunications equipment and their expected growth rates are set out in Table 4.2.

Table 4.2 Regional markets for telecommunications equipment

Region	1982	Projections $ billion (1979 prices)		growth rates (p.a) (%)
		1987	1992	
North America	19.9	29.1	41.9	7.8
Europe	12.5	17.2	23.7	6.7
Asia	11.8	19.1	31.7	10.1
Latin America	1.4	2.0	2.9	7.7
Oceania	0.9	1.2	1.5	6.6
Africa	0.4	0.7	1.0	8.2
Total ($ billion)	46.9	69.3	102.7	8.1

Source: Little (1983), cited in van Tulder and Junne (1984).

The telecommunciations market is heterogenous in products. The distinction is made between switching, transmission, terminal equipment and private systems. Table 4.3 shows the respective share of these different items. PTTs are the main customers, with 60 to 80 per cent of the purchases, depending on the country. Given the prominent role of PTTs, their procurement policy is the backbone of any industrial policy in each sector.

As far as trade is concerned, Europe has maintained a leading role, due to the absence of AT&T outside the United States since the 1920s. In 1963, Europe—the Six, the United Kingdom and Sweden —accounted for 83 per cent of world exports, while in 1976 this proportion was still 64 per cent. The leading firms are ITT, Ericsson and Siemens as was said above (see Table 4.4). The main losers over the last twenty years have been the British companies which have lost most of what was, in 1964, a significant (18 per cent) share of

Table 4.3 Telecommunications equipment manufacturing —the world market by type of equipment

	1976-80 %	1986-90 (projections) %
Terminals	9	7
Data transmission	2	4
Telex	3	3
Private switching (PABX)	12	14
Transmission	18	24
Public switching	56	48
Size of the market (yearly)	$32bn.	$68bn.

Note: Cables are not included.
Source: Sarati (1984).

Table 4.4 The twelve largest manufacturers of telecommunications ranged by sales in 1982 ($ billions)

	1973		1982	
	$ billions	Market share (%)	$ billions	Market share (%)
Western Electric (US)	5.67	40	13.01	33
ITT (US)	2.45	17	5.48	15
Siemens (FRG)	1.54	11	4.6	12
Ericsson (Sweden)	0.99	7	3.2	8
Alcatel-Thomson (France)	0.45*	3	3.1	7
GTE (US)	0.918	6	2.2	6
Northern Telecom (Canada)	0.51	4	2.14	6
Philips (Holland)	0.56	4	1.34	4
NEC (Japan)	0.72	5	1.28	4
Plessey (UK)	0.26	2	0.85	2
GEC (UK)	0.34	2	0.78	2
Italtel (Italy)	0.42	3	0.62	2

* CIT-Alcatel only.
Sources: Jequier (1976); Usine Nouvelle (1984).

the export market (Euroeconomics, 1978). But exports are only 10 per cent of world production. The oligopolistic structure of the equipment market was little disturbed by the bad performance of the British industry. The guaranteed home market is of more importance than export markets. Table 4.4 shows how little overall shares have changed. (Note that ITT and the European firms have grown faster than Western Electric reflecting the relatively fast growth of European markets during this period. These markets have now nearly reached saturation point for basic equipment.)

Western Electric, the AT&T subsidiary, remains by far the largest supplier of equipment. It is backed by the Bell Laboratories, and its record for innovation is unsurpassed.[3] Deregulation and the breakup of the AT&T monopoly in the United States mean that the company is now free to diversify both into foreign markets and into new (computer-based) products. Locksley (1983) estimates that by 1990 its exports will reach $6bn.

ITT, which grew from the rump of AT&T overseas subsidiaries hived off in 1925, is currently the world's largest exporter (see Table 4.5). The 1970s have been a bad period for the group. First came the débâcle in Chile. In 1976 it was forced to sell its French subsidiaries, Le Matériel Téléphonique (LMT) and Lignes Téléphoniques et Télégraphiques (LTT) and, in 1978, to reduce its participation in Standard Elektrik Lorenz (SEL) from 100 to 85 per cent. In 1975, it also cut its participation in Standard Telephone and Cables (STC) from 100 to 75 per cent. Its electronic exchange, System 12, was costly ($1bn) in R&D terms and now faces the competition of AT&T. In spite of this, ITT has maintained its market share surprisingly well, above all in Europe (Quatrepoint, 1984), particularly since it abandoned its diversification strategy and concentrated on its traditional strong point, communications.

Two companies which have done particularly well in recent years are Ericsson and Northern Telecom (see Table 4.4). The Swedish company, already one of the members of the leading oligopoly (together with ITT and Siemens), adapted itself quickly to technological change, thanks to strong co-operation with the Swedish PTT, with which it created a common research centre, Ellemtel. The Canadian company, Northern Telecom, is linked with Bell Canada, the largest network operation in that country. The key to its success is also a quick reaction to technological changes. It has benefited substantially from the deregulation of the American market where it made a massive entry, particularly in the PABX market. Together with Ericsson and Philips, it won a major contract in Saudi Arabia in 1978, beating an AT&T-led consortium. Northern

Table 4.5 Market share of ITT, Ericsson and Siemens in the world open market

	ITT		Ericsson		Siemens	
	% Market share	% Local production	% Market share	% Local production	% Market share	% Local production
Scandinavia (Sweden included)	15		50		10	
Austria and Switzerland	15		5		20	
Other Europe	40		10		15	
Total Europe (peripheral)		42		11		7
Oceania	25	24	10	6	7	9
Brazil	25		35		15	
Other Latin America	50		20		7	
Total Latin America		40		21		8
Total Developing Countries	32	38	18	15	13	7

Source: Aurelle and de Chalvron (1982).

Telecom licenses Olivetti of Italy and GEC of Britain for large PABXs (van Tulder and Junne, 1984).

Nippon Electric (NEC) is the largest Japanese company with 40 per cent of the home market. It has gained a strong position in semiconductors and computers, albeit less than Hitachi and Fujitsu, and is emphasising the concept of computers and communications —'C&C'. Other telecommunications manufacturers in Japan are Hitachi, OKI and Fujitsu. The Japanese industry gained a reputation for excellence in transmission equipment technology during the 1960s and 1970s, and has benefited from the transition to electronic exchanges to gain a foothold in the more important switching market. NEC has established a good position in the American deregulated PABX market, gaining a 5 per cent market share.

Siemens is by far the largest company in the EEC, its strength coming from its share in its home market, which is the biggest in Europe. It has traditionally also dominated the central European markets, although its position has slipped somewhat in the last two decades (see Tables 4.5 and 4.6). Its strength derives from the reliability of its equipment rather than technological flair.[4] Apart from Siemens, the leading firms in Germany are Standard Electric Lorenz (SEL), an ITT subsidiary since the 1920s; Telefonbau u Normalzeit (TUN), which belonged formerly to the AEG group, and Tekade, a Philips subsidiary which produces transmission equipment.

In the United Kingdom, the market is dominated by GEC, Plessey and, in a minor way, STC (until 1982 majority owned by ITT). Around the majors are some smaller companies like Thorn, the Ericsson subsidiary, Pye (a subsidiary of Philips), and Racal.

In France, the ITT subsidiaries, LMT, LTT and CGCT (Compagnie Générale de Constructions Téléphoniques) have been predominant for many years, but in 1976, the first two were bought by Thomson —at French government instigation—together with the Ericsson subsidiary in France (see below, Section III). The only French company to be involved in equipment from a fairly early stage was the Compagnie Générale d'Eléctricité (CGE), with its subsidiaries CIT-Alcatel and Cables de Lyon. Before 1976, Thomson produced transmission equipment only. Minor companies are the G3S group and TRT (Philips subsidiary). Matra is a new entrant, mainly for terminal equipment.

In Italy, the major manufacturer is Italtel, formerly SIT-Siemens, a subsidiary of the German company nationalised in 1945. Italtel now belongs to the STET group, controlled by the state holding company IRI. Italtel holds roughly 50 per cent of its home market. In addition there are the subsidiaries of multinational

Table 4.6 Switching market in Europe, 1974–1984

	1974	1984 (Digital switching)
Austria	Siemens ITT Local manufacturer: ITT patent	Siemens ITT 50% Local manufacturer 50% Northern Telecom Patent
Belgium	ITT 80% GTE 20%	ITT 80% GTE 20%
Denmark	Ericsson 70% ITT 10% Siemens 20%	Ericsson 80% ITT 20%
Finland	Ericsson 60% ITT 15% Siemens 25%	Cit-Alcatel (patent) 50% Ericsson 30% Siemens 15%
Spain	ITT 25% Ericsson 25%	ITT 70% Ericsson 30%
France	ITT 42% Ericsson 18% Cit-Alcatel 40%	Cit-Alcatel } 84% Thomson } CGCT 16%
United Kingdom	Plessey ITT GEC	Plessey } 100% GEC }
Greece	Siemens 40% ITT 40% Philips 15%	n.a.
Ireland	Ericsson 65% ITT 35%	Cit-Alcatel 40% Ericsson 40%
Italy	Italtel 50% ITT 20% GTE —5% Ericsson 15%	Italtel & Second pole
Norway	Ericsson 60% ITT 40%	ITT 100%
Holland	Philips 75% Ericsson 25%	Philips ATT 75%
FRG	Siemens 55% ITT 30% Tekade 15%	Siemens 60% ITT 40%
Portugal	ITT 50% Plessey 50%	n.a.
Sweden	Ericsson 100%	Ericsson 100%
Switzerland	ITT 35% Siemens 30%	n.a.

Source: Quatrepoint (1984).

companies: Fatme (Ericsson), Face (ITT) and GTE Italia. Two main outsiders are Telettra (of the Fiat group), which manufactures transmission equipment and has recently entered the PABX market, and Olivetti which has also made a breakthrough in PABX.

Overall, therefore, it is clear that the European industry has a fairly large number of manufacturers, but each national market is separated and protected, with close 'quasi-vertical' relationships between manufacturers and PTTs. The PTTs for their part were well aware that the quasi-vertical links which they had forged with manufacturers laid them open to 'opportunism' in the sense used by Williamson (1979), namely that the price of the equipment which they purchased would be too high and the quality too low. Various arrangements were made to avoid this problem. The British Post Office from 1928 until 1969 (and in effect until 1982) purchased by negotiated quota arrangements and co-ordinated R&D activities to make sure equipment came up to standard. (The Martlesham Research Centre of the PO (now British Telecom) leads research projects in co-operation with manufacturers and has been a major innovatory force—not always successfully.) In Germany much the same type of purchasing arrangements applied, but the dominant position of Siemens in effect created a bilateral monopoly and Siemens led most innovations, the DBP laboratory, FTZ, merely providing testing facilities for verification purposes. In France, until the 1960s, there was a bilateral monopoly situation between the ITT subsidiaries and the Direction Générale des Télécommunications (DGT), but this was broken in the 1960s by the entry of Ericsson and one or two smaller companies, including AOIP and CIT-Alcatel; these companies and the DGT together constituted SOCOTEL and SOTELEC, two 'clubs' respectively for exchange and transmission equipment which pool patents and co-ordinate research. In Italy, the situation is even more complex, since a previous system of local concessionaries has divided the country into zones in which one manufacturer enjoys a monopoly position as equipment supplier, with no co-operation at the research level.

The result of these protected and guaranteed home markets has been to promote inefficiency. Rugès (1970) estimated that the DGT paid at least 20 per cent more than the average price for its equipment. Recent figures are worse: an OECD study (1983) put the price differential between Europe and the United States at 60 to 100 per cent. The PTTs have tried to remedy the situation by introducing more flexible procurement methods and introducing internal competition among the national suppliers. But the major spur to flexibility has been innovation, and in particular the impact of electronics. It is this that we consider next.

The impact of electronics

Computers and semiconductors have been used increasingly in telecommunications over the last two decades. Computers are used essentially at the switching level; semiconductors for transmission (repeaters, multiplexers and amplifiers; also in terminal equipment). Satellites also use semiconductors extensively. Meanwhile, the diffusion of computers has increased the demand for data transmission facilities.

In order to understand the complex relationship between electronics and telecommunications, it is useful to distinguish between three aspects of innovation in the telecommunications sector:

(i) *Electronic switching*: which is a process innovation in the sense that it has not changed the final service itself (the transmission of voice) but has made it much easier and less costly, particularly for maintenance.
(ii) *Telematics*: which has come with the diffusion of the microprocessor and has brought product innovation in the form of a diversification of the service offered (no longer just voice communication).
(iii) *The ISDN* (Integrated Service Digital Network)—again a process innovation which builds on the two earlier developments to offer a completely transformed system. Messages are no longer sent via analog signals (frequency modulation) but digitally, which allows much more information to be passed, and thus the integration of voice with data and pictures. Ultimately, this development will transform telecommunications into videocommunications—although this is unlikely to happen before the year 2000, due to the large bandwidth requirement (more than thirty times a telephone channel).

Below we separate electronic switching from the other two developments. It merits special attention for two reasons: first is its central importance to the network; second is the fact that it is now basically over as an innovation, whereas the other two developments have yet to make a major impact on the system.

Electronic switching

There are two main parts to a telephone exchange. The *control part*, which monitors calls, identifies the caller, establishes contact with a free channel, holds the contact during conversation, opens it when it is finished and measures the number of pulses to be charged; and the *contact system*, which establishes the physical link with the system. Originally, the control was handled manually, usually by women who operated the control panel by plugging jacks into sockets. By the beginning of the century, automatic exchanges began to appear and, as we saw in the previous section, by the 1920s there were two leading systems, Strowger and AT&T's Rotary (which was in fact slightly superior) (Deloraine, 1974). After World War Two, the Crossbar system progressively supplanted the earlier two: as with Strowger and Rotary, the link is mechanical—two metal pieces make contact on command from an electric impulse—while the control is provided by electric circuits. Hence the name *electro-mechanical*, which applied to all these switching systems.

Electronic switching effectively replaces the electro-mechanical system by computer. The computer can be used, either to replace just the control function; or it can take over both control and contact functions. This distinction is reflected in the two types of technique. *Space division switching* uses computers for the control function but keeps the old physical link between subscribers during their conversation, usually through reed relays, which means that the signals transmitted are still analog signals (i.e. electrical impulses). *Time division switching* uses computers for both the control function and for translating the analog signals via pulse code modulation into digital signals. When combined with multiplexing, this allows for a substantial increase in system capacity. The important distinction is, however, between the analog transmission techniques of space division switching, and the digital techniques with time division.

The first attempt to introduce an electronic exchange was undertaken by the British Post Office, which began developing the exchange in the late 1950s at a time when computers still relied on valves! In spite of this, the aim was a time division switching exchange. The British manufacturers, although sceptical, went along with the Post Office initiative and the prototype exchange was opened at Highgate Wood in 1962. It was, however, ahead of its time: not until valves were replaced by semiconductors was this type of exchange to prove viable. In the meantime, the Bell Laboratories, where transistors had been discovered in 1948, had been undertaking research on space division switching using solid state devices, not valves, in its computer controls. The outcome, the ESS-1 was tested in 1964 and introduced into the network in 1966. Given its rapidly established reliability and cost advantages over time division switching, it is not surprising that the large manufacturers—Western Electric, ITT, Siemens and Ericsson—all chose to develop this technique rather than to pursue the risky path of time division switching.

However, for the European PTTs which had the research and financial resources to back the effort (namely the British and French PTTs), the risks and costs of time division switching had to be offset against the positive advantage it offered of enabling their manufacturers to leapfrog technologically over the established oligopoly: in effect it could form the basis of an industrial policy aimed at increasing the market share of their (national) equipment suppliers. For these same suppliers, to choose space division switching was to try to beat the established firms in their own field. Moreover, time division switching was a necessary step towards the ISDN—the dream of telecommunications engineers since the mid-1960s. Hence, from the network point of view, it had considerable technical attraction.

Electronic switching also provided the opportunity for reshuffling the established structure of the industry. As indicated in the first section, the only time the PTTs have the chance to redefine their relationships and/or introduce new suppliers is when a new generation of equipment is introduced. By the end of the 1960s, when it became clear that the decision over electronic switching was a crucial one, they had three alternatives:

(i) *to continue existing quasi-vertical relationships*, adapting to the new situation by working with manufacturers to develop a new product with co-operative R&D, pooling of patents, etc.;
(ii) to do nothing, *allowing time and competition to select the best system*, then encouraging national manufacturers to license this system;
(iii) *to try to modify the structure of the industry* to make it more suited to the new technological environment—for example by encouraging mergers or takeovers.

The first option clearly suited those PTTs which thought the existing structure of the industry could cope with the technological challenge posed by electronic switching; the second suited those PTTs which were anxious to take better advantage of their monopsony power; and the third suited those which had a clear view of the appropriate development of the equipment manufacturing industry.

Technical and institutional constraints meant that the choice was not wholly free. Nevertheless, it is remarkable that all of these options have been chosen at least once by the four major European PTTs.

CONTINUING QUASI-VERTICAL LINKS—EWS-A AND SYSTEM X

The story of the German and British attempts to introduce electronic switching via established manufacturers illustrates well the rigidities and difficulties which these quasi-vertical links entailed when the PTTs chose the minimal adjustment option.

The DBP decided in 1966 to phase out its old electro-mechanical exchanges. Given Siemens's monopoly over patents in switching systems, their space division electronic exchange, EWS-A, was the obvious choice. But when the first prototype was presented, in 1974, many experts doubted its success, given its bulky structure, doubts which were increased with the rapid development of microprocessors. Nevertheless, the DBP persisted with its plans, although Siemens itself began (after 1977) pursuing other research paths, while promising delivery of the EWS-A. It was not until 1979 that Siemens admitted the impossibility of meeting its promises. Meanwhile, the DBP had incurred some DM1,000m expenditures in infrastructure and manpower training (*Der Spiegel*, 10 September 1979).

System X was the British PO's belated response, after the Highgate Wood fiasco, to the need to develop an electronic switching system. For a time they shied away from it, preferring to concentrate on automating the physical connections rather than the control function (the TXE_2 and TXE_3 exchanges) and it was not until 1969 that they decided to look again at electronic switching. Their decision was to go for a full digital exchange based on the time division system and code-named System X. In effect, the PO's decision was a 'make or break' attempt to thrust the British industry back on the technological frontier.[5] The companies themselves were suspicious of the initiative: STC (the ITT subsidiary) was in any case involved with ITT's own electronic exchange, System 12, and inevitably had mixed loyalties (STC eventually pulled out of System X in 1982), and Plessey likewise had developed a small full digital exchange for the Ministry of Defence which the PO had refused in 1971. The end of the quota arrangements under the bulk supply agreement did not help matters: the PO had adopted the TXE_4 exchange (an enhanced version of TXE_3), developed by STC, together with the Crossbar system of Plessey and GEC. Two outsiders, Pye and Thorn-Ericsson, were ready to participate in System X manufacturing, although they had played no part in the R&D phase. Much time was lost in negotiations over who was to do what. Only in 1977, eight years after the original go-ahead decision, was the agreement reached: GEC would develop the computer, Plessey the switchgear and STC the control system. But even after the prototype was launched in 1979, disagreements continued.[6]

Overall, the story of System X is a dismal one. Far from providing a spearhead into export markets, British firms have been unable to get any sizeable order overseas, while British Telecom's (BT) ability to introduce new services is hamstrung by its lack of digital facilities. (see Table 4.7). The fault seems to lie in a failure of leadership. BT has the research capability, a staff of 2,450 at Martlesham and an R&D budget amounting to 3 per cent of turnover. But having taken the initiative in designing System X, it failed to follow it through— a factor not helped by the pressures for restraint on its investment from 1975/6 onwards. Manufacturers have also some responsibility for the over lengthy development of System X. As Professor Carter put it, 'it was not a natural team'; Plessey and GEC were afraid that the third manufacturer involved in the venture, STC, would pass its knowledge to the parent company, ITT. Moreover, the three producers have long earned good profits and employed a big work-force to build electromechanical equipment, even during the 1970s. They had no incentive to change—indeed it was to their advantage to prolong the life of the old exchanges as long as possible (Large, 1982).

As far as export orders were concerned, even its defenders admit that System X was designed primarily for the British market and was not geared to export markets. The result has been that the British industry lost the battle for a place amongst the prime suppliers of this generation of switching equipment and has probably forfeited

Table 4.7 British Telecom—exchanges in use, 1984

	Local		Trunk		International	
	No.	%	No.	%	No.	%
Strowger	3,708	59	331	72	—	
Crossbar	571	9	126	28	6	86
TXE	1,965	31	—	—	—	
Digital	48	1	2	—	1	14
Total	6,292		459	100	7	100

Source: British Telecom Prospectus, November 1984, Section E, 'Plant and Equipment'.

its place in the next generation (because it will be too small to meet the huge R&D costs involved). It also makes the British market very vulnerable to penetration by the telecom multinationals once the protection afforded by the quasi-vertical links with the manufacturers is removed. The recent decision of BT to buy, besides System X, other digital systems from Ericsson and Northern Telecom, is clear evidence of this.

THE COMPETITIVE OPTION—THE DBP ATTEMPTS TO RECOUP THE EWS-A FIASCO

Having announced the abandonment of EWS-A in January 1979, the DBP immediately declared that it was putting the provision of a new digital exchange out to competition: prototypes of the main contenders would be installed and subjected to a two-year test period between 1981 and 1983. Three contenders were chosen—Siemens with EWS-D; SEL with the ITT System 12, and Philips, whose subsidiary Tekade put forward the PRX system. (Two other systems —Ericsson's AXE and Northern Telecom's DMS were discarded because neither firm had manufacturing facilities in Germany.) In late October 1983 it was announced that the Siemens and SEL exchanges had been chosen.

It would, in fact, have been difficult for the DBP not to have chosen Siemens without destroying the German industry; and SEL also had 32 per cent of the German switching equipment market, so many jobs were at stake there too. But the experiment has given the DBP experience and knowledge of the current generation of electronic equipment and—most important of all—enabled it to reverse the old relationship with Siemens. It is now the DBP exercising its monopoly buying power rather than Siemens its monopoly selling power. As far as the latter was concerned, it had reacted quickly to the failure of EWS-A, putting 1,500 engineers onto the design of the EWS-D fully digital exchange and proving itself up to the two-year test period. Indeed, by the end of 1983 it had orders from twelve other countries for the same system. But the real gainer from the DBP's competitive strategy was ITT, which obtained the coveted DBP quality label for its System 12 exchange.

INDIRECT INTERVENTION: THE CASE OF PROTEO

In contrast to the other European countries, the telephone system in Italy is not run by the PTT, but by a concessionary, SIP (Società Italiana per l'Esercizio Telefonico). The PTT has a regulatory and planning function and operates only a part of the inter-city trunk network through an agency ASST (Azienda di Stato per i Servizi Telefonici). SIP belongs to the STET group, as does the manufacturer Italtel, and controls 80 per cent of the network, while ASST controls only 15 per cent. (The rest—satellites and cable—is given to two other STET concessionaries.) STET is in many respects similar to AT&T—it is both a major network operator and a manufacturer, but it is publicly controlled through the IRI. Hence, the government has two channels of influence over the telecoms sector in Italy: through the PTT with its minority network operator and regulatory function, and through the STET group. Not surprisingly, there have been clashes between the two of them.

In fact STET initiated research into digital switching as early as 1968 in an attempt to break away from the Siemens patents. The digital system, Proteo, was developed by a group of 450 engineers and scientists of SIT-Siemens (from 1980, Italtel), but from 1973 onwards successive financial crises (caused mainly by a refusal of Italian governments to sanction tariff increases), caused SIP to cut back investment. The result, therefore, of the first Proteo initiative was modest—a small PAM (Pulse Amplitude Modulation) digital exchange, introduced into the network in 1978. Conscious of their lagging technology, the Italian managers looked for American technology and installed a research unit in Dallas which produced an updated version, the Proteo UT 10/3 in 1982.

Meanwhile, the demands being made upon the network in Italy multiplied. In 1981, the PTT issued a ten-year plan giving priority to electronic switching, and announced in April 1982 that it was going ahead with two time division switching systems—Proteo and one other to be chosen from the multinationals operating in Italy—Ericsson, GTE and ITT. Italtel reached agreement with GTE to merge Proteo with the latter's EAX, and Telettra joined this team. Meanwhile, it appears that both Ericsson and ITT will get a share of the market. PTT policy in Italy has therefore indirectly forced a rationalisation of structure, but left the task only half completed. It also came too late. The continuing financial crises of SIP and Italtel throughout the years 1973-82 created gaps in network modernisation and development which will not be filled before 1990. This makes Italy, like Britain, vulnerable to entry by the new international majors, AT&T and IBM.

DIRECT INTERVENTION—FRANCE

While other European countries refrained from direct intervention in the telecommunications industry, the French government stepped in, twice, within a period of seven years, to effect a major transformation of the French equipment industry.

The background to this development stems from the DGT's work on electronic switching. Its laboratory, CNET (Centre National d'Etudes des Télécommunications—today numbering 4,500 researchers) embarked on time division switching in 1965, working with a small company—SLE-Citerel, a subsidiary of Ericsson and CGE. The outcome of this co-operation was the first fully digital time division exchange in the wrold, the E 10, which began to be installed in the French system as early as 1972. (It was a very small system and suitable mainly for rural exchanges.) Unfortunately, subsequent co-operation between CGE and CNET for a larger exchange did not work so well.

The first intervention came in 1976 when, with the announcement of a 140bn frs ($29.2bn) investment programme for the years 1975-80, the DGT, under the pretext of discussing the provision of space division exchange systems, manœuvred ITT into a position in which it was forced to sell a 75 per cent stake in its subsidiaries to Thomson, a group which until that time had interests only in transmission equipment. Subsequently, Thomson also took over the Ericsson subsidiary in France, thus acquiring within a short period a 40 per cent stake in the switching equipment market. CGE (CIT-Alcatel), the other main French supplier of equipment, was left largely untouched by this restructuring. These moves had effectively 'Frenchified' the industry—which until then had been dominated by multinationals—and in line with the Giscardian philosophy, created two competing national champions, one (Thomson) having an expertise in space division, the other (CGE) in time division. Both benefited from the massive investment programme of the years 1975-80.

The second round of restructuring came in 1981 with the new Socialist Government. Thomson and CGE were then both nationalised, and ITT's remaining subsidiary, CGCT, was bought in July 1982 by the French Government. Then, in September 1983, came agreement to merge the telephone activities of the two French groups and to bring them successively under the control of CGE (and taking the CGE name, CIT-Alcatel—by now it was clear that time division switching was the way forward to a full digital system). In line with Socialist principles, one strong nationalised group has been created, capable of intervening in the market economy (see Boublil, 1976) and of carrying the French flag in international markets. But such a rapid structural transformation also entails costs. Thomson, even in the earlier period when expansion was still rapid, faced problems of integration and overmanning. With the fall in the investment programme and the emphasis increasingly on technological developments—many of which are capital-intensive but labour-saving—these problems multiplied. Moreover, with AT&T and IBM both anxious to establish a presence in the European telecommunications market, it is not clear that even one unified equipment manufacturer in France can survive. A. D. Little, the management consultants (quoted in *Il Mondo Economico*, 1983) estimate that in order to recoup R&D costs, a switching equipment manufacturer has to reach 3 per cent of the world market. Alcatel and Thomson together should have over 6 per cent, but they still have two systems to develop.

ELECTRONIC SWITCHING: CONCLUSIONS

What conclusions, then, can be drawn from these various attempts to cope with the challenge posed by electronic switching? First, the dilemma between space division and time division was really a non-issue. The dominant firms had all the structural and brand image advantages from their position in the electro-mechanical technology. Embarking on the most reliable and profitable new technology, namely space division switching, was the natural innovation path for ITT, Siemens, Ericsson and Western Electric, *provided* they all did the same thing. The resulting situation is called a Nash equilibrium in oligopoly or game theory. For outsiders, time division switching was a good option, offering the chance of capturing the lead at the technological frontier. The two most successful companies today, Northern-Telecom and Ericsson, won their reputation for excellence with digital switching, although at the end of the 1960s both embarked on space division switching. But they were quick enough to react at the right time (the mid-1970s) to the transition. Britain, Italy and France, which had all embarked early on time division switching, should have been better placed to take advantage of it.

Two contradictory statements provide some explanation of the poor performance of the European telecom industry. Diodati (1980) argues that the quasi-vertical links were not strong enough between manufacturers and PTTs. Others emphasise the necessity for a more competitive procurement policy. (See, for example Babe (1981), who shows that vertical integration leads to overpricing.) In a sense, both are right. If time division switching was chosen, vertical links needed to be tight, given the risks and costs of the venture. If space division was preferred (as for example in Germany), competition would have given the necessary incentives to shift technologies when it became clear that the future lay in the other technique and avoided the opportunism (in the Williamson (1979) sense) of the dominant supplier.

In the countries embarking on time division switching (France, Italy and Britain), the objectives of PTTs and manufacturers were well aligned in the early 1970s. Demand and investment were high, which made co-operation easier. But the cutback in investments from 1974 onwards (apart from in France) created a divergence of objectives: manufacturers were anxious to maximise production of electromechanical systems in order to keep current plant operating (and avoid problems of redundancy and lay-off), and they had no incentive to put resources into the development of new electronic equipment. In contrast, Northern Telecom and Ericsson, although benefiting from research co-operation with Bell Canada and the Swedish PTT respectively, had no guaranteed home markets. In order to maintain their position in international markets they had to innovate.

The outcome of the electronic switching episode has been to leave the European industry in a relatively weak position in the face of international competition. Technological advance now means that the expected life of exchanges is ten years rather than the thirty to fifty years of Strowger and Crossbar, while the R&D cost has (in relative terms) risen. Table 4.8 gives details of the R&D costs for the respective electronic switching systems. The range is from $500 to $1,400m. Annual sales revenues necessary to recoup these costs need to be at least fourteen times this amount, yet the largest European market, Germany, has annual sales of only $11,000 (EC Commission, 1983).

This puts in stark relief the quite undue fragmentation of European industry. Yet the experience of the 1970s on restructuring was not

Table 4.8 R&D cost of digital switching systems ($ billions)

System 12 (ITT)	1.0†
AXE (Ericsson)	.5 ‖
E10 and E12 (CIT-Alcatel)	1.0†
DMS (Northern Telecom)	.7*
System X (GEC/Plessey/BT)	1.4*
ESS-5 (Western Electric)	.75‡
EWS-D (Siemens)	.7§

* *The Financial Times*, 24/25 August 1983.
† *Usine Nouvelle* 14 June 1984.
‡ *Telephony* 25 April 1983.
§ Personal contacts
‖ *Il Mondo Economico*, 31 March 1982.

good. Admittedly, the German problem was to break the dominant position of Siemens (which it did by introducing competition), but for other countries the problem was that manufacturers were too small and too dependent upon PTT protection. It would seem that direct intervention coupled with a high level of demand (as in France between 1975 and 1981) is the most successful strategy, but the period of fast expansion is now over and even if other countries were prepared to take a leaf out of the DGT's book, conditions are no longer ripe. Moreover, national markets remain too small for a 'national champion' strategy.

The trend, therefore, has to be joint ventures and technology transfers among manufacturers. Already one sees the emergence of a number of US–Europe links—the GTE, Italtel, Telettra grouping in Italy; AT&T and Philips have joined forces to merge AT&T's ESS 5 and Philips's PRX digital systems. However, there are obvious dangers in becoming too dependent on American technology: it opens European markets to American multinationals but can make European firms unduly dependent on foreign licensing for employment and profits. The alternative must lie with European co-operation—but whether this is best achieved by orchestrated collaboration or market forces is a moot point to which we shall return later.

Electronic switching has given the European PTTs the opportunity to help their suppliers leapfrog the dominant manufacturers. This experiment, which can be considered as a genuine industrial policy, has failed, with perhaps the exception of France. Today, R&D costs for developing a new generation of digital exchanges, combined with the shorter life expectation of this equipment, creates the necessity (rather than an opportunity) to build a new industrial policy, this time at the European level. In other words, the task of keeping pace with technological progress in the 1980s seems far more difficult than in the 1970s.

From telecommunications to videocommunications
—the revolution in network provision

Telematics and videocommunications, unlike electronic switching, are product innovations which have altered (and are altering) relationships both with end-users and with manufacturers. Moreover, the pressures on the telecommunications sector have been not only *technological* but *institutional*. Both pressures have been very much intertwined during the last ten years, but the PTTs have sometimes reacted better to one than to the other.

The technological pressures can be roughly divided into three phases. The first originated from the increased transmission capacity of the networks, which came with statellites, microwaves and high capacity coaxial cables, combined with increased demand stemming from the diffusion of computers and time-sharing techniques. The use of modems—the means by which digital computer messages are translated into analog signals for the telecommunications network—multiplied five-fold in Europe between 1972 and 1979. (See Table 4.9 for 1981 figures.) The telephone network, given that it remained an analog system, was not well suited to this new, fast rising source of demand, although it managed to cope with it. PTTs leased dedicated lines to large-scale data transmission users (e.g. banks), or created specialised data networks, with circuit or packet switching. But they saw these as temporary measures, necessary until they could digitalise the whole network. Again, it reinforced their long-term objective—the ISDN.

Table 4.9 Data transmission terminals in Europe, 1981

	Up to 1,200 bits/sec	From 1,200 to 9,600 bits/sec	Over 9,600 bits/sec	Total
France	53,390	44,434	737	98,561
West Germany	62,496	64,686	125	127,307
Italy	55,444	29,709	133	85,286
UK	67,994	19,898	347	88,239

Source: Eurodata Foundation (1981).

The second phase of technological pressures came with the advent of microprocessors—which meant that terminals could become 'intelligent'. The main developments are as follows:

(i) *Videotex*—combining the TV and telephone and offering the subscriber an interactive service in which he can consult data bases, make financial transactions, book travel, etc;
(ii) *Teletex*—a service somewhat like Telex (but much quicker and more flexible) in which messages can be sent via word processors;
(iii) *Facsimile*—which enables documents to be copied via the telephone network and therefore sent from one location to another very fast and without using the postal services;
(iv) *Electronic Mail*—this is a service which combines teletex and facsimile, to provide subscribers with 'electronic mail boxes';
(v) *Teleconferencing*—two or more people can be brought together

by videophone or telephone only (audio-conferencing), for a conference or meeting with exchange of documents, graphics, etc., via facsimile transmission;
(vi) *Mobile telephone*—this service, which was hampered by the band width of the frequency spectrum, was boosted by a more rational use of this band width, developed as a result of the use of microprocessors.

These services were developed as add-ons or 'value-added' services primarily for the business market, although with spare capacity on residential lines, the PTTs have been anxious to develop telematics in this sector as much as in the business sector. (See chapter on videotex.) Here to a degree lay a conflict of interest, since the business sector, with its increasing demand for data transmission, felt that investment should go into network development rather than services for domestic markets. This conflict was really between the public service function of the PTTs—to provide an equal access for everyone to service the telecommuncations—and a marketing approach, which emphasised the necessity of quickly recouping innovative but expensive investments. Needless to say, the PTTs were more accustomed to the first than to the second task.

The third phase of technological innovation has been videocommunications, either through the enhancement of ISDN or through cable television. The latter has been hindered from development in Europe by regulatory factors, but in the United States there were 20 million subscribers in 1976, although a cost-effective interactive technology was not available until the 1980s. The development of optical fibres, with a capacity many times that of coaxial cables, has boosted plans for 'intelligent' cable TV (CATV) in Europe, although optical fibres are currently an expensive option.

The institutional pressures on the PTTs were just as strong as the technological ones. They arose from the deregulation of the American market where, until 1983, AT&T had dominated 80 per cent of the network (but had been restrained from competing in other sectors—e.g. computers—and on international markets). Its monopoly position was attacked by degrees. First came the Carterphone decision in 1968 which liberalised the terminal market; in 1971 the Specialised Common Carrier decision allowed competitors to enter the field of data-transmission in competition with AT&T; in 1972, entry was permitted to satellite transmission; in 1976, the specialised common carriers were allowed full interconnectability with the AT&T network (which allowed them to compete in voice as well as data transmission). Then, in 1982, came the decision to divest AT&T of its twenty-two local companies, but in return AT&T was freed from restrictions on its own activities. This meant AT&T could compete in the computer and office automation sectors, and in international markets.

Developments in the United States inevitably had an impact on Europe: governments could no longer duck the issue of innovation in this sector. With demand for data transmission, though growing fast, still low by American standards, the PTTs were able in the first phase to respond by reiterating the ISDN concept. Their monopoly over the common carrier network was not really questioned, partly because of trade union strength in that sector. But with the development of telematics, conflict between business and PTT interests became greater and were quickly translated into the political arena. Demand for data transmission facilities was growing very fast, while the PTTs themselves seemed intent on 'empire building' into mass telematic services. But it was difficult to decide on priorities because, by their nature, many of these new services needed to be supplied before the market could decide whether they were viable. (Again, this is an issue explored in the videotex chapter.) With the third phase—cable TV—it is the telephone network itself which is in question, since the cable network represents an alternative method of providing voice, data and videocommuncation. But it raises the issue of whether there is any longer any point in providing the PTT with public monopoly status.

Thus, the technological and institutional debates have developed side by side but, as time has gone by, the institutional debate has become more urgent. The following discussion looks at the answer to these issues arrived at by the individual countries and tries to assess whether their strategies had any validity.

GERMANY—THE DPB'S ATTEMPT AT MAXIMAL PRESERVATION OF MONOPOLY

In the early 1970s the German telephone network was the largest and most modern in Europe. It was also in some senses the most rigid. In the previous section, we have detailed the degree to which until recently Siemens dominated the switching equipment market: but the same hegemony applied to much of the terminal equipment market where the DBP had, via its regulatory controls, effectively protected the position of the major German manufacturers, particularly Siemens. This situation has in many senses been transformed over the last decade: more competition has been introduced while, at the same time, the DBP has updated the network and the services available. Nevertheless, the DBP has remained firmly in control throughout. Pressures for more radical changes, involving, in particular, the removal of the DBP's monopoly over regulatory and transmission facilities, have been gently but firmly ignored.

The blueprint for these changes came from the KtK report (Kommission für den Ausbau des technischen Kommunikationssystem) in 1976. It mapped out the future development of the network, suggesting a rapid expansion of transmission facilities, and the implementation of some narrow band telematic services (facsimile and electronic mail). But the bulk of investment was to go on traditional telephone services, with no initiative for broadband networks (i.e. cable) apart from a very limited testing programme.

The KtK report was very much the product of the first phase of technological development and mainly reflected the DBP's view that no institutional change was necessary. In 1977–9, the DBP made a number of strategic errors. First was the EWS-A fiasco; but it also cut back on investment while maintaining high tariffs, allowing the profits to help finance the deficits of postal services and the Federal Budget. This triggered off a bitter political debate which not only questioned the wisdom of the decision over broadband services, but opened up the whole question of public versus private operation.[7]

The reaction of the German Government and the DBP has been to defend the network monopoly while opening up procurement. We have seen how this happened in the electronic switching decision, but the same occurred in videotex, where in 1982 IBM won the contract for installing the system in competition with SEL (ITT's German subsidiary) which had done preliminary work in this area for the DBP. Mitel, a small Canadian company producing PABXs, gained DBP approval for its digital SX/100 private exchange. As far as broadband services were concerned, an experiment was announced in 1981 with financial help from the BMFT (Federal Ministry for Research and Technology) for the provision of local videocommunication services using optical fibres (Project Bigfon). It involves six rival projects, with newcomers such as Krone, Quante and Nixdorf, alongside the traditional Bundespost suppliers. The cost of the Bigfon programme is to be $135m, of which half comes from the BMFT.

At the same time, the DBP has resumed investment in the telecommunications network, and attempted to align prices more closely to costs. New connections in 1980 were 1.6 million, compared to 700,000 in 1976. The Bundespost was the first PTT in the world to introduce teletex in 1981 (it was a Siemens innovation) and priced it cheaply in order to attract customers quickly. It also opened a new packet switching network in 1981 for data transmission, and a new network for the mobile telephone was also opened, increasing capacity to 100,000 subscribers. Cable TV was further boosted with the change of government in 1982, with investment reverting to the traditional coaxial cable, but raised to $300m. This means that current investment by the DBP is the highest in Europe.

Nevertheless, there is still much criticism of the DBP. Firstly, in spite of opening up its procurement policy, the DBP is still considered by experts as the least liberal PTT in Europe (Locksley, 1983, p. 109). Many examples support this opinion. Nixdorf launched a digital PABX in the United States two years before it attempted to do so in its home market; the 'network C' mobile telphone is not considered to represent 'state-of-the-art' technology; in other areas, the high standards of the DBP are held to be an obstacle to exports (*Communications International*, 1983). Secondly, the EWS-A fiasco has delayed digitalisation of the network, which is far behind the French, and has thus delayed the transition to the ISDN. The DBP is accused of excessive enthusiasm for videotex (see Chapter 5), while the money going into expanding the cable TV network (with the full monopoly of the DBP as common carrier) is being put, it is claimed, into an obsolete technology (*Wirtschaftswoche*, 1984). Finally, it is claimed, the dominant position of Siemens has hardly been challenged, while the recent attempt to set up a cartel around Siemns for the provision of optical fibres has not been condemned by the DBP, although it has been declared illegal by the courts.

The change of government in 1982 did not bring the institutional change the Christian Democrats had been asking for in opposition. The DBP's strategy throughout has been clear. After initial vacillation (up to 1979), it has followed the three waves of innovation—data transmission, telematics, and cable TV—with a high level of investment, simultaneously manœuvring to keep its monopoly position. The risk lies in attempting to combine these three initiatives—it is not yet clear that the DBP's plans or structures are flexible enough to cope. In the meantime, equipment manufacturers benefit in two senses: on the one hand they gain from the high levels of investment, while on the other the DBP monopoly guarantees the survival of a *German* industry.

FRANCE—THE DGT'S BRIEF HOUR OF GLORY

Telematics in France date from 1978, which was the year of the seminal Nora–Minc report[8] which preached the central importance of the telephone network to information technology and the dangers of domination by IBM. The year 1978 was also in the middle of the massive investment programme in telecommunications in France. The DGT managers who backed the Nora-Minc proposals for a major state-backed initiative in telecommunications were equally conscious of the need to provide the equipment industry with new markets once the 1975-80 investment programme was over.

The outcome of these pressures was the 'Plan Télématique' of 1978, which involved a series of initiatives involving both network development and a wide range of terminal telematic services. As far as the latter was concerned they comprised:

(i) *a videotex experiment* of the French system (Télétel) at Velizy, a suburb near Versailles, which involved 2,500 voluntary subscribers;
(ii) an experiment involving an *electronic directory* which provided the inhabitants of St. Malo with free, but compulsory, use of electronic directory information. Each household was provided with a cheap terminal with VDU and keyboard for accessing information—equipment that could also be used for Télétel (videotex);
(iii) a norm was established for *facsimile transmission* and subsidies given for the development of a cheap facsimile terminal for residential use;
(iv) an experiment with *electronic money*, testing point-of-sales payment methods.

On network developments, two major experiments were put in hand:

(i) the development of a *communications satellite*—Télécom 1 (to rival IBM's SBS)—aimed at the business market and bringing such services as video conferencing and high-speed data transmission;
(ii) an experiment in *optical fibres* offering broadband facilities, including cable TV, high-speed facsimile transmission, videophone etc. The experiment was to be based on Biarritz and involved 1,500 subscribers—the largest test of its kind in the world.

The year 1979 also saw the completion of the Transpac venture, a special network developed with packet switching and dedicated to data transmission. The total cost of all these measures is difficult to estimate, but according to the DGT itself (see *Télécommunications*, hors-série, October 1983), the expenditure for telematic services only (excluding the satellite and the Biarritz test) would amount to some $400m by 1986.

The provision of these services marked a major departure for the DGT. Until this date its main function had been laying cables and looking after the networks created. The new range of telematic services meant that for the first time the telephone service was seeking to create new markets. (Until this time it had never had to go out and sell its services.) But the market repercussions were not lost on the DGT. It was conscious that after 1980 manufacturers would be looking for other outlets: moreover, a healthy home base in telematics provided a good springboard into export markets. This was the rationale behind the electronic directory: by making its use obligatory it immediately created a protected home market for 30 million VDUs! Unlike the DBP which had launched a series of experimental tests to choose the new equipment, the DGT chose suppliers through negotiation. This brought in some new entrants— e.g. Matra and TRT (the Philips subsidiary in France)—but generally the DGT remained firmly in control, spearheading a vertically integrated relationship with manufacturers and focusing on wide industrial goals as a contrast to the traditional PTT concern with network development.

The boldness of the DGT's strategy was greeted with fascination but also with scepticism. As far as the manufacturers were concerned, they behaved as subcontractors rather than as self-conscious oligopolists, acquiescing to the DGT's lead, protesting at the difficulty of making terminals cheap enough to meet specification, but generally reacting rather passively. The DGT's urging to go out and find export markets fell on deaf ears. The Socialists, taking political advantage of the situation, argued for a more comprehensive strategy to restore French competitiveness in the computer and components industry. But when they came to power in May 1981, they in fact went along with most of the telematics initiatives: the satellite Télécom I was launched in June 1984, the Biarritz test was opened in 1983, and France is now being supplied with Minitel electronic directory terminals, although they are no longer compulsory. (See chapter on videotex.) Only the residential facsimile experiment has been abandoned—on economic grounds.

Socialist policy took the industry in other directions. First, CGE and Thomson were nationalised in 1981. Then, in 1982, a comprehensive plan for the *'filière éléctronique'* was launched, which gave no central role to telematics (whereas the Giscardians had tended to place it centrally within the *filière*), but did require the DGT to finance two-thirds of projected government support to the *filière éléctronique*. This support amounted to $1,800m a year (*Le Boucher*, 1984). To meet this requirement, tariffs had to be increased, and the DGT became in effect a milch-cow to the *filière*, and also the general budget. The DGT meanwhile meekly commissioned studies of office automation, computers and semiconductors —somewhat of a reversal of roles.

A new orientation was also given at the network level with emphasis on two directions:

(i) narrow band ISDN to overlay and enhance the data transmission network Transpac, enabling simultaneous voice and data transmission. This network, due to be opened in 1987, is aimed at the small business user;
(ii) development of integration of narrow and broadband services via cable. In November 1982 a comprehensive plan for building cable TV networks with fibre optic cables was announced, with 20,000 connections in Montpellier to be used for the first commercial trial. Meanwhile, the DGT is pushing ahead with plans for 1.4 million subscribers to be connected up to fibre optic networks by 1985, with an overall investment of $1.6bn for the years 1984-7, 60 per cent of this coming from the DGT. (There are, however, difficulties with optical fibre supplies. Only one small firm, Velec, in joint venture with CGCT, gained DGT approval for its product on first-round trials.) (Betts, 1984.)

There has never really been debate in France about the monopoly role of the DGT. Interventionist policies have accumulated, one upon the other, but the DGT has always been looking ahead and never really questioned where it was. There is, however, a risk of overload amongst the accumulated tasks which could lead to institutional crisis.

As for the equipment industry, it has performed well enough, failing none the less to meet the ambitious goals set by the DGT. New entrants like Matra, which is making cheap VDU terminals for Minitel, after being encouraged by the DGT to enter the market, were left out of the bulk orders and had to look overseas for markets. The major incumbents, CGE and Thomson, meanwhile, continued to get the bulk of the orders and although this seemed to go along with some success in export markets (sales up from $600m in 1978 to $1.2bn in 1982), recent figures indicate some slackening, particularly in 1984. There is little doubt that the future of the industry in France is closely tied up with the success of the DGT in its common carrier role. The latter is likely to suffer, the greater its ambitions to go beyond this role.

ITALY: PRIVATISATION THROUGH THE BACK DOOR

Mention has already been made of the financial crisis which dominated Italian telecommunications from 1974 to 1981. One of the consequences of this crisis was constraints on SIP's ability to keep up with technological innovation. In none of the three aspects—data transmission, telematics or cable—has SIP really played much part. Now it lags behind the other PTTs in all these add-on services, as well as facing a telephone penetration rate that is low by European standards. There is now, therefore, substantial competition between investment to expand the basic network and investment to provide for new services.

In the 1970s, experts in Italy were very conscious of the need to develop a modern telephone network and that this required state aid. The Communists supported the ASST (the PTT regulatory organisation which also runs 15 per cent of the network, including important trunk routes) against the STET group, which they accused of 'playing the multinational game'. The SIP for its part was anxious to introduce new services, but could not do so without the approval of the ASST to tariff increases by which to meet the expenditures. But the ASST was anxious to extend its own network by taking on services such as those proposed by SIP. There were inevitable conflicts of responsibility and the outcome was that nothing was achieved—a somewhat typical example of the Italian bureaucratic disease.

Things began to change in 1980 when the tariff issue was resolved. The long-awaited PTT programme was issued in 1981, but was far too cautious—for example, investment in data transmission services was scheduled to be only 8 per cent of investment. The aim was to reach 22 million subscribers by 1990—a level surpassed in Germany in 1981! Within the programme were plans for the launch of videotex, a packet switched network for data transmission, satellite and teletex services. Videotex began in 1982, on an experimental basis while circuit and a packet switched service (Itapac) came in 1983. Both services are provided by Italtel with FACE (ITT) as subcontractors. A satellite is planned for 1987.

Within this perspective, the multinational subsidiaries seem to have only limited involvement with telematics. Olivetti, a powerful potential entrant, having tried a joint venture with a French group, Saint Gobain, has, since December 1983, formed a strong alliance with AT&T which has got 25 per cent of its shares. The Italian company will sell the American products in the European markets. IBM has also moved into the field, proposing to STET that it help run the packet switched and circuit switched networks, in exchange for the choice of its own communication software, SNA.

On the institutional front, the concession to SIP which ended in 1984 was renewed during the summer of that year. SIP was given a wide mandate for new services but the long-awaited fusion with the state agency ASST did not happen. At the same time as this new concession was issued, SIP was allowed to increase tariffs, but the discussions, begun with IBM the year before, did not end with the expected agreement on a data network. Instead it was agreed that work on components and robots, produced by the STET subsidiary SGS–ATES for IBM, should proceed on a common basis. IBM has still, therefore, to obtain a proper foothold in Italian telecommunications.[9]

In many respects, Italy combines all the disadvantages of the American regulatory regime *and* the European model of government control. This creates inherent conflicts of interest at the political, rather than the strategic level. As a result, Italy is now too far behind to catch up, and it is not surprising that the two major American firms, AT&T and IBM, are both trying to enter the market. In many ways it is a preview of what could happen to Europe, if it does not succeeed in implementing a successful telecommunications policy. It is only with a dynamic and innovative PTT, that the technological challenge can be coped with.

BRITAIN: IS DEREGULATION THE PANACEA?

In the United Kingdom the reaction to telematics has become dominated by one issue—that of institutional control. In many senses the British Post Office was one of the earliest authorities to recognise the potential that developments in computers offered the telecommunications network. As we have seen, both in Highgate Wood and in System X, it made bold bids to put the United Kingdom and the British industry on to the leading edge of switching technology and, in the early development of Prestel, it recognised the exciting potential of marrying up with computers to create an interactive service. It is too facile to explain failure in terms of these ideas being 'ahead of their time'. Admittedly, the technological problems of electronics switching had to wait upon the development of the semiconductor, but time and again the Post Office and the British equipment industry failed to capitalise upon the clear technological advantage which they at one time held. The story of Prestel (see chapter on videotex) is just the latest episode in a history of poor management and industrial squabbling.

The issue of the institutional framework really goes back to 1969, when the Post Office became a public corporation (as distinct from a Department of State) and was required to separate the accounts of its telecommunications activities from its postal and girobank activities. As a public corporation, it had greater autonomy over its management decisions, but its investment plans were subject to government approval and were financed through the central exchequer (i.e. it could not issue debt or shares to the Stock Exchange). In 1976, Professor Charles Carter was asked to review the workings of this new public corporation and his report (UK Government, 1977) made a firm recommendation that the two activities, posts and telecommunications, be separated, and each created into a public corporation in its own right. The main focus of his attention had been on electronic switching and the unsatisfactorily slow progress of System X. He felt that the link with the postal services, with their traditional, slow-moving pace of innovation, hindered the flexibility necessary in the fast-moving sector of telecommunications. He urged the rapid introduction of System X and the liberalisation of the terminal equipment market (until then monopolised by the Post Office) in order to secure a rapid expansion of new services. He did not focus much on telematic services, or upon data transmission, and he left the institutional question untouched.

The Carter Report led directly to the establishment of British Telecom (BT)—the nationalised industry for the supply of telecommunications services—which was set up in 1981. But the new Conservative Government simultaneously set up a review (under Professor Beesley) of its monopoly status, and as a result of this review introduced measures in 1981 aimed at liberalising the telecommunications markets both in terms of equipment sales and, more significantly, in terms of service provision. Mercury, a joint venture between Cable and Wireless. Barclays Bank and BP—was licensed in 1982 to build a new, fibre optic inter-city service network aimed primarily at business traffic.[10] At the same time plans were announced to 'privatise' British Telecom through a share flotation which would sell off 51 per cent of the enterprise to private investors. A new regulatory agency was to be established and British Telecom was to become, like AT&T, a private monopoly (or to all intents and purposes a private monopoly) subject to regulatory control.

Although institutional questions dominated the adjustment process, the Post Office/British Telecom also tried to keep pace with the technological developments. Data transmission requirements from the early 1970s onwards were dealt with via leased lines and modems. A packet switched network called PSS was introduced, but not until 1981; Prestel was the hallmark of the telematics strategy (see Chapter 5) and an increasing number of specialist services are now on offer via Prestel. A teletex service was opened in 1984, while facsimile services were developed in 1983/4.

Nevertheless, the threat of competition (via Mercury) has stimulated BT to formulate some sort of strategy. The introduction of System X has been accelerated, and a small prototype of narrow band ISDN called IDA (Integrated Digital Access) was set up in London in 1983 to accelerate the transmission of substantial data sets (e.g. clearing between banks). The use of optical fibres in trunk transmission has been emphasised, tariffs (overall) were stabilised but individually aligned more closely with costs, and the waiting list for new telephones (260,000 in 1980) reduced to 3,500. Mercury, meanwhile, were given interconnection rights with the BT system, in spite of opposition from the Post Office Engineering Union. Mercury also faces start-up difficulties and does not seem to be going to offer services which in any way enhance BT's provision, while their tariffs will be the same. The British Government has blocked other network entry for the next five years to give Mercury a chance to establish itself but, all in all, it would not appear that Mercury really offers much threat to BT.

BT faces potential competition from two other sources: cable TV and the mobile telephone. Provision of the latter went to public tender in June 1982 and approval was given to two competing systems, one from BT, the other from Racal-Milicom. In cable TV, the United Kingdom has followed the American pattern of putting local franchises up for public tender. Eleven franchises have now been awarded, with BT being a partner in roughly half of these. But the prospects for cable TV in the United Kingdom are still uncertain.

The equipment manufacturers in the United Kingdom have greeted liberalisation and privatisation with suspicion. They claim to seek 'fair play' and fear that BT would use its dominant position for its own benefit. Perhaps more to the point, they fear the entry of newcomers into the British market. The issues are well illustrated by the situation in the PABX market where BT enjoyed a monopoly in the provision of small systems, but did not compete at all in the provision of large systems. After deregulation, observers expect BT to get 50 per cent of the large system market in 1986/7 (de Zoete and Bevan, 1984), while retaining its 65 per cent of the small system market. BT currently sells GEC/Plessey equipment or Mitel's, depending on customer preference, but newcomers Ferranti (with GTE), Harris, Thomson, CIT-Alcatel and Siemens are knocking at the door.[11]

The policy of the Conservative Government and its emulation of the American deregulation model has dramatically changed the structure of telecommunications in the United Kingdom. Nevertheless, many European features still remain. While it is true that competition for terminal equipment supply is now well and truly open, and there could be competition in other value added services, the main problems of BT—an outdated network and overmanning—are not really susceptible to competition from Mercury. It has been the *threat* of competition rather than its implementation that has been important. It has pushed ahead the (very necessary) internal restructuring of BT along functional lines (until this time it had retained its old civil service organisation) and brought tariffs more closely into alignment with costs. In this respect it is arguable that privatisation *per se* is irrelevant to the objective of improving efficiency, and that liberalisation, bringing with it the threat of competition, would have been sufficient. Had the amount of management time and energy consumed by privatisation been devoted instead to the strategic management and development of the system (*à la* DGT), British Telecom would today be a far healthier company and British industry better equipped to face the future of the information revolution.

The future: can the PTTs survive?

Looking back over the evolution of the telecommunications sector in Europe in the last two decades, we have noticed the increasing importance of technology and institutional change. Today, however, the industry faces a new challenge—the potential entry of IBM and AT&T—which seem to have chosen Europe as one of their battlefields. Assessing the chances of survival of a telecommunications industry in Europe will thus require an assessment not only of technology and institutional issues, but also of the new industrial challenge.

The technical progress of telecommunications through the three phases discussed in this chapter has brought out the dilemma currently faced by telecom authorities—diversification or integration. If they go for the first, then they need competition, enabling them to provide a wider range of services with flexible market strategies and competitive pricing systems. This, broadly speaking, is the option chosen by the United States. With integration, on the other hand, the unity of the transmission network is paramount. Digitalisation becomes the technical prerequisite for such unity—although this is not incompatible with the diversification route.

It is difficult to say whether one option is superior to the other. What is clear, however, is that *they do not have the same time path*. Diversification decisions can be taken quickly, as in the United Kingdom, and bring quick positive results—as in the United States. The integration route, by contrast, is slow because it requires a significant part of the newwork to be digitalised before it can offer interactive services which will attract customers. On the other hand, in the end it has a good chance of offering a superior service *because* digitalisation is in itself an advantage.

The answer given by most European PTTs to this dilemma has been clear for some time: they opted for the ISDN and hence integration. The reason for their choice was that the ISDN preserves economies of scale while giving the full range of service. In some senses this was the typical engineer's vision, focusing on technical performance—the ISDN was a major improvement on the analog telephone system—and long-run network planning. What was lacking was any notion of how to deal with demand in the transition phase. In fact demand as a whole has always had low priority for the telecommunications engineers; they were brought up to expect it always to exceed supply.

In theory, the European PTTs have followed a common pattern in planning and implementing their ISDN, although the pace has obviously varied. It can be described as follows:

First — Specialised leased lines or dedicated circuits as the answer to the first technological challenge (the need for data processing capacity). This was the PTTs' response up to the end of the 1970s.

Second — Experiments with mass market services as an answer to the second (telematics) phase of innovation: this preoccupied the PTTs at the end of the 1970s and the beginning of the 1980s.

Third — Narrow band ISDNs which will be introduced from 1987 to 1988 onwards, bringing together policies (1) and (2) above.

Fourth — Complete digitalisation of the telephone network and diffusion of local cable TV networks, which will come in the 1990s;

Fifth — Videocommunicatons through a broadband ISDN will merge steps (3) and (4) and will not be available before the twenty-first century.

Although this pattern has some coherence, given the decision to adopt the integration route, its practical application leaves much to be desired. Each country has deviated, in one way or another, from the model.

In Germany, it looks as if there will be a variety of competing networks, each run by the monopolistic Bundespost. In a document issued early last year (see *Deutsche Bundespost*, 1984), a qualitative picture of networks by 1992 shows four independent sources of transmission:

(i) direct broadcasting satellite connection between TV stations and private customers;
(ii) TV coaxial cable connections with the same kind of service;
(iii) telephone cable connection for business and residential users, giving narrow band services like facsimile, teletex and videotex through the telephone network;
(iv) optical fibre connection for business users with needs for large bandwidth services such as videoconferencing, high-speed data transmission and videotelephone.

One gets the impression that investment is scattered over all these competing networks, delaying the break-even point of each one.

In France, the situation is even worse. Two satellites provide telecommunications service: Télécom I, already launched, gives high-speed transmission for business users, while TDF 1, to be launched next year, will offer direct broadcasting facilities. Besides this, the DGT has, as we have seen, plans for a narrow band ISDN service, a large cable TV programme, and it is providing residential users with a cheap and downgraded videotex terminal, giving access to telematic services through the existing telephone network.

In Italy, the major issue is not the excess, but rather the lack of plans. Indeed, investments are insufficient to meet the pace of technological advance and end user needs. Diversification will perhaps impose itself *de facto*, as a consequence of this lack of public entrepreneurship.

Already by 1980, the United Kingdom Post Office, albeit slowly by European standards, had had to plan for ISDN. Now liberalisation and privatisation have confused this option. BT, as the dominant player in the sector, has still the capability to pursue this strategy, but short-run profitability considerations and the presence of Mercury will induce the company to behave along the lines of AT&T or MCI (the major AT&T challenger in the United States). This means that digitalisation at the local level could be delayed while big business users are provided with highly profitable services.

The European PTTs have thus mixed elements of the diversification strategy with the integration route, thereby weakening it considerably. It seems likely that by trying to keep control of the whole sector, they will not be able to keep up with the pace of innovation necessary for a fully diversified (deregulated) route. In other words, the main problem with the European PTTs is not that their stubborn defence of monopoly is old fashioned and economically unsound, but that this defence, combined with their desire to provide all kinds of services and networks at more or less any cost, will weaken their position. Customers of the traditional telephone service will not be prepared to subsidise indefinitely the costs of too many risky and mutually damaging projects. France and Germany, in particular, seem to have accelerated the third, fourth and fifth phase of development, while the second phase still has to show some benefit.

As far as Europe is concerned, the monopoly debate, and more generally the institutional questions, are not the main issues. Looking across the PTT experience of the last ten to fifteen years, it would appear that the PTTs which have performed best (France and Germany) were least troubled by the institutional questions, while Italy and the United Kingdom have had their investment and innovative capacities hampered by the debate over institutional issues. This does not mean that this debate has been useless or harmful (witness what was said in the previous section about improvements in BT competitiveness) but it has prevented the PTTs, at least temporarily, from focusing on the technological transition to the ISDN (i.e. from modernising their networks).

The pros and cons of an ISDN (integration) strategy can be summarised in the following way. Conceptually, the ISDN has three major drawbacks:

(i) Although the PTTs remain a strong element in the telecommunications sector, their importance, in terms of market share, is diminishing. The transmission network and the service it provides represented 78 per cent of the total market for telecommunications services in 1980 but by 1990 this proportion will fall to 69 per cent (Gerybadze and Friebe, 1983). The fast-growing markets are 'value added' ones, and terminal equipment. PABXs, for example, will see their share rise from 10 per cent to 20 per cent by 1990.

(ii) There is still some uncertainty about how the new services delivered through the ISDN will be used. Although this problem is common to any telecom strategy, a diversification framework allows the winners to be picked more easily, while the centralisation involved in ISDN creates the risk of rigidity. Users will be tied to a 64 Kbits/sec. transmission.

(iii) There will continue to be the same problem of network provision as exists today, for example cross subsidisation between profitable and loss making users. This hampers innovation by diverting resources to other objectives.

On the other hand, the ISDN has at least three major merits:

(i) With narrow band ISDN (step three of the general pattern above and the main focus of PTTs for the next ten or fifteen years) there is no need to recable the networks, and most of the demand of residential and business users will be satisfied. Moreover, digital transmission will bring greatly improved image resolution.
(ii) It reduces uncertainty for terminal manufacturers who know in advance the norms and likely market developments, and can adjust their production capacity accordingly.
(iii) ISDNs can be standardised at the European level quite easily, which brings at least two advantages: first, it enables a common policy approach to be adopted, and second, it would give the European PTTs a strong bargaining position, since no such standards exist at the moment in the United States. The discussions held by the CEPT (Conference Européenne des Postes et Télécommunications) on a common ISDN standard, though not enforceable at the national level, are a first step in the right direction.

The diversification option also has its drawbacks:

(i) It cannot avoid the creaming-off issue, where entry of competitors in profitable markets leaves the incumbent running the loss-making services. In the deregulated American market, the seven Bell operating companies have preserved their monopoly after AT&T's divestiture, partly for this reason, and each of these companies runs, on a pure monopolistic basis, a network as large as any of the national networks in Europe.
(ii) Deregulation is not always a synonym for quick innovation, because in a competitive framework private benefits of R&D are not necessarily equal to the social ones (Dasgupta and Stiglitz, 1980). Innovation is quickly imitated, thereby diminishing the expected return of an R&D effort.
(iii) The threat of entry by large American companies like IBM and AT&T at the service provision level should not be underestimated. The risk is that the large companies impose *de facto* standards,

and thus enjoy a rent not justified by their technical excellence. The suggested link between IBM and BT rang the alarm bell in this respect.

All in all, the integration option seems better suited to European conditions than the diversification option: first, because, as British experience shows, it is actually difficult to create competition in network provision; secondly, because the PTTs are, if given sufficient autonomy, quite capable of implementing a dynamic and innovative strategy; and finally because, given the current state of the European equipment manufacturers, European governments cannot afford to allow free entry to the large multinational companies.

Opting for the integration strategy does not mean that monopoly must be preserved at all costs. Terminal equipment and some value added services could be liberalised, as in the United Kingdom, without threatening the overall strategy of the PTT; indeed, such liberalisation enables them to concentrate attention and resources on their main, common carrier, role. What is important if the integration strategy is to succeed, however, is that the full ISDN pattern is introduced gradually, along the lines set out at the beginning of this chapter, and the PTTs do not attempt to accelerate procedures, as some have been doing.

Integration, of course, provides no answer to the problems of the equipment manufacturers. This is a more complex issue and will be dealt with in the next section.

A European scenario

If integration via the ISDN is the choice of most European countries, it is not a European choice. Would a common policy improve the position of the European PTTs? The European Commission thinks it would and is trying to set it up.

Its efforts have three main thrusts. First, to create a common market for telecommunications equipment, which means eliminating non-tariff barriers, opening up procurement policies to European bidders, getting the PTTs to agree upon common standards and accelerating approval procedures for new equipment; second, to eliminate the present duplication of R&D, which is becoming more and more costly. The most striking example of this is electronic switching where there are five European systems (E10, MT20, System X, EWS-D and Proteo), compared to one in Japan and two in the United States; third, to promote some common projects as a catalyst for common action by the European PTTs. One such project has already been implemented—Euronet/Diane, a scientific data base accessible from everywhere in Europe—and a second is in the pipeline—INSIS, a narrow band ISDN to connect Commission offices in Brussels, Luxembourg and Strasbourg.[12]

Commission officials maintain that a common European telecommunications infrastructure is as important today as the Steel and Coal Community was in the early post-war period. There are, however, difficulties, not least over design, which has become a hot political issue between countries, with protracted debate at Council level and PTTs reluctant to yield sovereignty.

In any case, the European PTTs are getting together bilaterally under their own auspices. Most of the initiatives to date have come from the French, and have not met with much success. A suggested exchange of digital exchange orders with BT failed because the French were not prepared to build a plant in the United Kingdom (Marsh, 1985), and a project between the DBP and the DGT to adopt a common procurement policy for the cellular mobile telephone has been postponed until 1990 (Marsh, 1985).

The PTTs are also playing the 'US game'. The STET-IBM agreement on semiconductors and robots was the first step, and the BT-IBM proposals on data transmission networks (blocked by the British Government on anti-trust grounds) would have been a significant major second step. At present, the PTTs seem more inclined to deal with an American partner than another European partner, while the Commission is trying to persuade them to do the reverse.

Manufacturers have also been looking for partners, again, generally with American companies. Mention has already been made of the Philips and Olivetti deals with AT&T, the GTE-Italtel link and the IBM-STET agreement. But Ferranti has GTE patent rights for PABXs and Siemens has the Xerox patent for the Ethernet local area network. The world leader in fibre optics, Corning Glass, licenses the major European manufacturers for its products. These deals form part of a much broader movement in the fields of computers, semiconductors and communications with companies jostling with one another to find the fittest partner for an appropriate venture. There is a general attempt to find complementarity: computer manufacturers search for a partner with communications interests and vice versa. It is much more difficult to link up with companies with similar interests.[13]

The fact is that the American firms often have more to offer in the way of complementary interests (too many European firms are too similar in interests). This helps to explain why there is so little collaboration at a European level. To reverse this would require more than political will—rather, it would require:

(i) self-sustained technical and economic evolution;
(ii) better perception of the issue at a political level;
(iii) greater automony and a more coherent policy for the PTTs; and
(iv) some kind of 'federative agent'.

As far as the first, technical and economic evolution, is concerned, this is likely to show that the most profitable network development for the next fifteen years in Europe is narrow band ISDN. This will be linked with other developments in office automation (local area networks and PABXs), integrating voice and data transmission.

The second factor is political. In effect, co-operation at a European level for some projects will require an act of political will: more discussion of the common market will not be enough. There are analogies between telecommunications and air transport with state monopolies offering captive markets to manufacturing industry. The political commitment (and the choice of a marketable product!) was a necessary condition for the success of Airbus. The same is likely to hold true of telecommunications.

The third factor, greater autonomy for the PTTs, is also vital. To date, the financial effects of government intervention have been very damaging, milking investment resources, tying them up with administrative rules and isolating them from their markets. If competition is introduced for the value added services, the PTTs will have to have greater financial flexibility than at present. Closer market contact by the PTT would also provide a major spur to efficiency for the equipment manufacturers.

Finally, if co-operation is to work at a European level, it needs some kind of cohesive or 'federative' agent. The DGT has tried to play this role since 1983, without much effect. Generally, the DGT is much weaker today than five years ago. Given the size of Siemens in the world market and the resources it can command, the Bundespost is better placed today to take the initiative, even though its relative lack of R&D expertise is a weakness.

Even if all the conditions for the successful implementation of a European policy were met, there remains the question of a marketable project. Here the options are wide—ranging from the next generation digital exchange, through fibre optics, to sophisticated terminal equipment such as 'intelligent' PABXs and local area networks. What is needed is to choose the project which best fits the narrow-band ISDN strategy which would underlie the European initiative.

Telecommunications is one of the very few activities in electronics where Europe is stronger than Japan and can compete on equal terms with the United States. But this position is fast being eroded. The strategy outlined here of integration via the ISDN, plus European co-operation on specific projects, plus liberalisation at the margin, offers a chance to reverse this erosion. Whether Europe will follow the route of pessimism or optimism has yet to be seen.

Notes

1. In fact Bell registered his patent only two hours before Gray tried to register his patent and it has long been a moot point as to which one was the first inventor.
2. This made the telephone system the first example of nationalisation in Britain. Unlike subsequent nationalised industries (e.g. the BBC, National Coal Board, etc.), it was not set up as a separate corporation with its own accounts, personnel, etc., but was run as a department of government, responsible directly to a minister of the crown, the Postmaster-General, and through him to Parliament. His employees were civil servants and paid as civil servants and its accounts were consolidated with those of the Exchequer. This remained unchanged until 1969, when the Post Office became a public corporation embracing both post and telecommunications. In 1981 two separate public corporations were established for the two services, the Post Office and British Telecom. There is one small segment of the telephone service which has remained outside the nationalised system this period—the municipal telephone system in Hull.
3. Many of the technical breakthroughs in telecoms come from the Bell Labs: feedback amplifiers in 1927, transistors in 1948, gigahertz microwave systems the same year, waveguides in 1936, Rotary and Crossbar switching systems in 1919 and 1938 respectively, electronic switching, at the experimental level, in 1947 (see Libois, 1983).
4. For example, its EMD system which was patented in Germany in 1955 used a technology analogous to the Rotary, implemented by AT&T just after the First World War.
5. The term 'make or break' comes from the Carter Report (1977). After twenty years of bad technological choices reflected in loss of international market share, the Post Office attempted in System X to push through a major innovative effort which would, had it succeeded, have taken the British companies into the forefront of the new technology.
6. Locksley (1983) maintains that there was disagreement even over pricing strategy. STC wanted to price low in order to gain market share—even to the point of losing money in the early years, while GEC wanted a price which would recoup the development costs.

7. Mettler Meibom (1983) argues that the cautious attitude of the KtK was a compromise between manufacturers and regions. The former did not want interactive wideband CATV which could compete in the long run with their EWS-A analog system. The latter feared the one-way coaxial CATV was a threat to their control of TV and radio programmes. Hence, the CATV option was discarded.
8. Nora and Minc, *L'Informatisation de la Société* (1978). This report, which appeared at the beginning of 1978, awakened both government and public to the importance of the convergence between computers and telecommunications. The authors coined the neologism 'télématique' and warned against the challenge from IBM through the satellite ventures SBS which would offer high-speed business transmission facilities. The report has been a bestseller and translated into various languages.
9. During the first half of 1985, the STET group seems to have changed strategy. It is trying to keep the full control of network provision, on a monopoly basis, while participating through its minority shareholding in initiatives taken by the two 'majors'—Olivetti (with AT&T) and IBM—in value-added services. It remains to be seen whether this strategy is sustainable or whether the multinationals, having once got a foothold in service provision, will not wish to enlarge their presence.
10. Cable and Wireless is a British company specialising in network installation and management in the open markets. It was denationalised just after the Conservatives were returned to power in 1979; Barclays resigned from the Mercury project in 1983, and British Petroleum left in summer 1984.
11. The new freedom given to BT after privatisation (November 1984) has been welcomed by the company's management. But the relations with the British Government have become ambiguous. A deal with IBM for a value-added service network was rebuffed by the Government on antitrust motives. In the Spring of 1985, BT acquired a majority shareholding in the Canadian PABX manufacturer, Mitel. This second deal has been referred (by the Government) to the Monopolies and Merger Commission which must consider whether it enhances BT's monopoly position before approval. Moreover, the regulatory agency, OFTEL, does not seem to have the capabilities really to control and monitor BT's behaviour. BT, for example, rejected outright its suggestions that it should limit its non-System X purchases to 20 per cent of new exchanges.
12. A third major initiative, the RACE programme, has been adopted in March 1985 and is based on the earlier ESPRIT programme. It involves a partial financing by the Commission of precompetitive R&D projects involving co-operation of several European manufacturers (IBM being considered as European). The aim of RACE is to lead, from 1995 onwards, to the creation of a wide bandwidth trans-European network—as part of the basic infrastructure for information technology. The French EUREKA proposals could also have implications for telecommunications networks and technology, as, of course, does the US Strategic Defence Initiative.
13. In November 1984 four major equipment manufacturers—Plessey, Italtel, Siemens, and Alcatel-Thomson—announced that they would in future be

producing in common the customised integrated circuits (chips) used for switches in their electronic exchanges. This co-operative initiative, although notable after years of natioanlism, is nevertheless still very modest, and far from creating a European telecommunications industry as claimed by some newspapers.

Bibliography

Aurelle, B. and de Chalvron, J. G. (1982), 'Le secteur des télécommunications: destabilisation et restructuration', *Bulletin de l'IDATE*, No. 8, September.
Babe, R. E. (1981), 'Vertical Integration and Productivity: Canadian Telecoms', *Journal of Economic Issues*, March, pp. 1–31.
Betts, P. (1984), 'Why France is taking such a bold gamble', *The Financial Times*, 15 May.
Boublil, A. (1976), *Le Socialisme Industriel*, PUF, Paris.
Le Boucher, E. (1984), 'Premiers résultats positifs', *Le Monde*, 23 November.
Brock, G. W. (1981), *The Telecommunications Industry*, Harvard University Press, Cambridge, Mass.
Cherry, C. (1977), 'The Telephone System: Creator of Mobility and Social Change', in I. de Sola Pool (ed.), *The Social Impact of the Telephone*, MIT Press, Cambridge, Mass.
Commission des Communautés Européennes (1983), 'Télécommunications: Communication from the Commission to the Council', Com. (83) 329 final, 9 June, 13 pp., Brussels.
Commision des Communautés Européennes (1984), *Report from the Commission to the Council on Telecommunications*, Com. (84) 277 final, Brussels.
Commissione Morganti (1984), *Rapporto al Presidente del Consiglio sulle Telecomunicazioni*, Franco Angeli (ed.), Milan.
Communications International (1983), '*Special Report-FRG*', August.
Dasgupta, P. and Stiglitz, J. E. (1980), 'Uncertainty, Industrial Structure and the Speed of R&D', *Bell Journal of Economics*, pp. 1–28.
Deloraine, M. (1974), *Des Ondes et des hommes*, Flammarion, Paris.
Diodati, J. (1980), *Vertical Integration as a Determinant of Industry Performance in the Telecommunications Industry*, European Association for Research in Industrial Economics.
Eurodata Foundation (1981), *Yearbook*.
Euroeconomics (1978), *Telecommunications in Europe: The Next Ten Years*, Brussels.
The Financial Times (1983), 'World Telecommunications', Part I, 24 October. 'World Telecommunications', Part II, 25 October.
Gerybadze, A. and Friebe, K. P. (eds) (1983), *Microelectronics in Western Europe: Medium Term Perspective, 1983–87*, Berlin, Eric Smidt-Verlag.
Hoare and Govett (1984), *British Telecom*, prepared by R. Pringle, P. Roe and R. Gray, London.
Jequier, M. (1976), *Les Télécommunications et l'Europe*, Centre de Recherches Européennes, Lausanne.

Kommission für den Ausbau des technischen Kommunikationssystem (KTK) (1976), *Telekommunikationsbericht*, Dr Heger Verlag, Bonn.
Large, P. (1982), 'Three companies into one project won't go', *The Guardian*, 5 April.
Libois, L. J. (1983). *Genèse et croissance des télécommunications*, Masson, Paris.
Little, A. D. (1983), *World Telecommunications Program*, Wiesbaden.
Locksley, G. (1983), *The EEC Telecommunications Industry. Competition, Concentration and Competitiveness*, Evolution of Concentration in Competition, Series No. 51, Commission des Communautés Européennes, Brussels.
Marsh, P. (1985), in 'International Telecommunications', *The Financial Times*, 15 January.
Mettler-Meibom, B. (1983), 'Versuche zur Steuerung des technischen Fortschritts', *Rundfunk u Fernsehen* No. 1, January.
Mondo Economico, Il (1983), 'Rapporto Mese: Telecomunicazioni', 31 March.
Nora, S. and Minc, A. (1978), *'l'Informatisation de la Société'*, Le Seuil, Paris.
OECD (1983), *Telecommunications: Pressures and Policies for Change*, Paris.
Quatrepoint, J. M. (1984), 'Cit-Alcatel n'a fait qu'une percée limitée sur les marchés européens', *Le Monde*, 15 May.
Rugès, J. F. (1970), *'Le Téléphone pour tous'*, Le Seuil, Paris.
Sarati, L. (1984), 'Lo Sviluppo delle reti di telecomunicazioni nel mondo' in *Telecomunicazioni*, No. 2.
Télécommunications (1983), hors série, *La DGT et la recherche en télécommunications*, pp. 50-8, October.
Telephony (1983), 'An Inside Look at French Telecommunications', 25 April, pp. 26-33.
UK Government (1977), *Report of the Post Office Review Committee* (Carter Report), HMSO, Cmnd. 6850, London.
Usine Nouvelle (l') (1984), 'Télécoms, la chance de l'Europe', No. 24, 14 June, pp. 42-8.
van Tulder, R. and Junne, G. (1984), *'European Multinationals in the Telecommunications Industry'*, mimeo., Universiteit van Amsterdam.
Williamson, O. E. (1979), 'Transaction Costs Economics: The Governance of Contractual Relations', *Journal of Law and Economics*, 22, pp. 233-61.
Wirtschaftswoche (1984), 'Management by Moses', No. 36, pp. 36-44.
de Zoete and Bevan (eds) (1984), *British Telecom*, London.

5 Videotex: much ado about nothing?

Godefroy Dang Nguyen and Erik Arnold

Videotex is the name given to a range of techniques which involve the transmission of information from a computer to the television screen. This chapter is concerned with one particular part of this wider family—interactive videotex—which allows the information recipient to respond by passing messages back to the information provider (to the computer), which can in turn be acted upon (for example, booking an airline ticket). To do so the recipient needs to be linked back to the computer transmitting the information, and normally this is achieved via the telephone line. Throughout this chapter the term videotex will be used to denote the more specific interactive videotex.

Interactive videotex was invented by Sam Fedida, working at the British Telecom (then the Post Office) laboratory at Martlesham, in 1971. Its development was in fact a by-product of experimental work on picture telephony (Fedida and Malik, 1979). It was based on the idea of using a simple modification of two widely used consumer durables—the telephone and the television—to provide domestic consumers with cheap access to computer data bases. By using the existing telephone network and domestic television sets, the hardware costs of the new service could be minimised (i.e. consumers did not have to buy a special dedicated VDU or computer; nor did the linkage require a special cable network), and this was seen as a major inducement to take-up. However, the use of the existing telephone network implied relatively slow data transmission rates and a narrow bandwidth, while the use of television as a display technology imposed a low definition picture standard 40 characters wide (as against the 64 and 80 character standards used in professional VDUs).

For the major 'actors' on the supply side, videotex offered a useful diversification. These actors were:

(i) the PTTs* who operate the telephone network;
(ii) the computer industry supplying computers for switching and storing data bases;

---
* Throughout this chapter, in line with the chapter on telecommunications, we shall use the term PTT to designate the public telephone network authorities—British Telecom in the United Kingdom, the DGT in France and the DBP in West Germany.

(iii) television or terminal manufacturers, who saw it as stimulating demand for a new type of television set;
(iv) information providers (IPs) who supply the data bases—banks, mail order companies, travel agencies, press groups, even the civil service.

Private videotex systems, such as the one operated by travel agents in the United Kingdom, can be incorporated into the public system via the so-called closed user group procedure, although they can also be implemented by the direct linking of private computers.

It would be possible to draw several different diagrams of the interrelationships between these actors in terms of inputs and outputs. In the most usual model, the innovating PTT designs or acquires (and possibly modifies) software, buys computers to implement this software, imposes standards, and acts as the vendor of videotex services. Sometimes this includes supplying the terminal, as the PTT has traditionally done with the telephone handset. Alternatively, videotex could be offered by a computer manufacturer on its own machines. This has been attempted in a limited way by GEC in the United Kingdom, but this model was not a success and other computer manufacturers have not followed suit. Finally, IPs themselves may exploit their knowledge of potential subscribers' information needs to organise a videotex system, as in the Reuters City Service.[1] In this last case, a range of alternative but similar technologies has been available from the computer manufacturers, and is used in such applications as airline bookings, news services and financial information (about, for example, commodity prices). The choice to implement a service using the public videotex channels rather than some other form of videotex represents simply a choice of one amongst a number of alternative existing techniques, but it takes advantage of the capacity of the telephone network to provide, immediately, a huge potential market.

The concept of the *filière*

This web of relationships between the actors can be described by the term *'filière'*, now common in French industrial economics. The word *filière* has been taken over by politicians and journalists to such an extent that its meaning is often now obscured. According to Toledano (1978), it originates with work on 'triangulating' economic input–output matrices which permits the identification and ranking of inter-industry flows of goods and services so as to highlight degrees of interrelation between industrial sectors. Analysis may be done on the basis of inputs or of outputs. Thus, where inputs are considered, if industry C buys 40 per cent of its inputs from industry A, while industry B buys 10 per cent, then A and C are considered to be more closely related than A and B. Chains of such important relationships may be charted, beginning with primary goods and ending with consumer goods and services. Generally, this chain revolves around a strong technological relationship (e.g. *filière éléctronique, filière bois* . . . , etc.).

A *filière* therefore involves a flow of goods through a succession of primary, intermediate and final goods sectors. Movement between sectors involves a market which may or may not be internal to firms, depending on the extent of vertical integration. Analysis of goods flows along a *filière* and through these successive markets has obvious uses, particularly in devising industrial policies which are designed to exploit the interrelatedness of certain sectors to provide multiplier and other effects. Policies may be chosen which, for example, minimise the extent to which there is dependence on international trade. Such policies may also involve intervention by government to capture control of one of the parts of a *filière* which dominates the performance of the whole—for example, semi-conductors are often seen as having strategic significance within the micro-electronics *filière*.

The concept of the *filière* is logically separable into two categories: the *macrofilière*, which is the notion of the *filière* as a chain of relationships leading from primary to final goods and the *microfilière*, which comprises one link in such a chain together with its relationships to the links immediately up- and down-stream. For example, electronics represents the *macrofilière*, while semiconductors represent a *microfilière* feeding into the *macrofilière*. This is in some senses very close to the usual idea of an industry or sector, but the term '*microfilière*' expresses the interdependence of the growth of the products involved, and hence of the economic actors. The conventional idea of a sector, in contrast, involves a single market and a single product or group of products. Figure 5.1 shows the distinction between *micro-* and *macrofilière* diagramatically.

In terms of the development of a *filière*, both Malsot (1980) and Mistral (1980) suggest that in the early stages of production the control of the *filière* and of its evolution is located upstream, whence the main technological characteristics of the product stem. Soulié, however, has argued that this is only likely where product cycles within the sectors involved in a microfilière develop in step (Soulié, 1983). In the initial stages of a product's life, technical aspects are

Videotex: much ado about nothing

	Macrofilière	Microfilière
Primary sector	⊤	1
		1
		1
Intermediate sector	1	1 2
		1 2
		1 2
Intermediate sector	2	1 2 3
		2 3
		2 3
Intermediate sector	3	2 3 4
		3 4
		3 4
Intermediate sector	4	3 4 5
		4 5
		4 5
Intermediate sector	5	4 5
		5
		5
Final goods sector	⊥	5

Figure 5.1 *Macrofilière* and *Microfilière*.

not well mastered and the characteristics of demand for the new product are not well understood. Those who have introduced the technical innovations involved in the product are better placed than those who market it to adapt it to demand, if only through a process of trial and error. We would see this advantage dissipating in favour of imitators and exploiters of the original technical breakthrough, as they accumulate technological capability through working with the technology and as they gain knowledge about the character of demand through the marketing process and through seeing the mistakes of the first-comers.

But even when a product becomes mature, firms continue to adapt. Malsot (1980) has argued that process innovations tend to increase the number of production steps until a new and radically different way of producing is discovered. This may establish a new *microfilière*, with its own pattern of growth and development. During the period of incremental improvement, control of the *filière* tends to migrate downstream in the direction of final demand. This is likely to be true both of the degree of control exercised by actors on the supply side and the control over the product specification exerted from the demand side: it is far easier for consumers to imagine minor improvements to existing products than to conjure up and specify to firms on the supply side what their demands might be for radically new types of products. The downstream movement of control within the *microfilière* as it matures may be expressed through vertical integration by upstream firms, as they try to retain control. Equally, the downstream shift in power can provide new strength to firms located there, which may themselves exploit this by upstream integration. A radical innovation can, however, side-step this maturation process through the creation of a new *filière*. As a result, the strategies of entrant and incumbent firms must be redesigned.

Videotex as a *microfilière*

In this sense, videotex can be seen as a *microfilière*, even to the extent of being an attempt by the PTTs to introduce a radical innovation which bolsters their position within the telecommunications *macrofilière*. It comprises the following upstream/downstream relationships:

(i) the computer industry which sells hardware to the network providers;
(ii) network providers, who sell telecommunications services and/or data base capacity to IPs;
(iii) IPs who in turn sell information to subscribers (this last relationship is muddied by the mediating role here of the PTTs as providers of infrastructure and often as selling agents of the IPs);
(iv) terminal manufacturers, who may supply end-users directly or sell to the videotex system supplier, who acts as intermediary.

In practice, the policies of the PTTs have determined the shape of the web of interrelationships within the videotex *microfilière*. Figure 5.2 shows the slightly different shapes of these interrelationships at the inception of the public service in the United Kingdom, France, West Germany and Canada.

The importance of the software used to implement videotex was not lost on the PTTs, who realised early that the standards embodied in videotex software were a potential source of control over the *microfilière* world-wide while, at the national level, the control of these standards could bring advantages to other parts of the telecommunications and electronics *macrofilères*. This, combined with their existing monopoly of telecommunications wiring and their accumulated technical know-how in telecommunications, meant that the PTTs straddled the new *microfilière* of videotex, necessarily

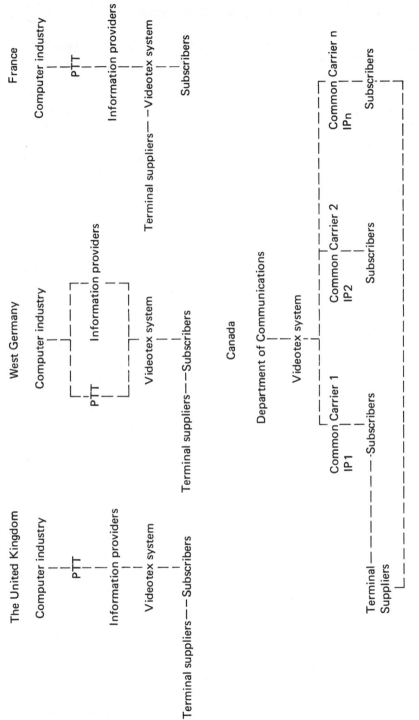

Figure 5.2 The interrelationships (*microfilières*) within the various European and Canadian videotex systems.

ensuring that they would have played a crucial role even if they had not also been the source of the technological innovation itself.

Better than any other actors, the PTTs were in a position to provide the infrastructure for videotex—not least because they already dominated the provision of telephone services through the telephone network. They were able to spread the risks and costs of innovation over a very large number of 'shareholders' (telephone subscribers), and videotex appeared to be a natural extension of their services. Given the approaching saturation of European telephone markets and the low rates of usage of domestic telephone lines, PTTs saw videotex as an important opportunity to extend their product range and increase revenues at relatively little marginal cost. But the PTTs also saw videotex as an important link into the electronic *macrofilière*. It gave them (or so they thought) an important lever within their own electronics industries in terms of purchasing computers, VDUs, etc. The leading PTTs involved in the United Kingdom, France and Canada also attempted to sell their own version of these standards abroad so as to establish a competitive edge for members of the electronics *filière* in their respective countries. Thus, videotex was seen as a potential method of exerting a far wider influence over the complex of electronics industries than initially seemed likely.

As we shall see, these aspirations were exaggerated. To date, videotex as a *microfilière* has proved to be relatively unimportant: far from controlling technology at an important 'nodal' point of conjuncture, it has seemed something of a dead end, and one that has been rapidly overtaken by developments in computer, cable and satellite technology. In retrospect, this is perhaps not surprising. In technical terms, it was a very minor innovation—so 'obvious', for example, that British Telecom were unable to patent any aspect of its invention.[2] What is surprising is that so many people should have built such high expectations on so minor an innovation.

This latter point deserves further consideration. Basically, the PTTs thought that since the *potential* market for videotex was huge, the *actual* market would be huge. They underestimated the price-elasticity of demand and the length of time it would take consumers (end-users) to get accustomed to the idea. They also underestimated the software difficulties (both their own and those of the IPs) and they failed to appreciate the fact that some IPs saw it mainly as an extension of advertising systems. But it is a matter for consideration whether the major error of the PTTs has not been on timing rather than anything else. There may actually be a need for a cheap, mass orientated, easy-to-use data communications system. Microcomputers —the obvious alternative to videotex—do not completely meet this need. The merging of the two—microcomputers and videotex— may provide the answer. It may well be that the advent of the cheap home computer will provide the incentive which has so far been lacking to develop the potential of videotex.

Videotex in practice: experience in the United Kingdom, West Germany and France

THE UNITED KINGDOM

Sam Fedida first demonstrated what he called a 'viewdata' system in 1973 at British Telecom's* Martlesham research centre. Like AT&T in the United States, British Telecom had experimented with, and rejected, video telephony in the early 1970s. It was clear that useful moving pictures could not be transmitted through the narrow bandwidth of the existing telephone system, and the capital cost of recabling was not then countenanced. Faced with low usage of domestic telephone connections and a desire to increase revenue, British Telecom's position was clear. 'The opportunity arises in the provision of new services that can be accommodated on a telephony bandwidth channel. It is then possible to provide the new service quickly over a wide area without massive capital investment.' (Post Office Research Review, 1978). It was of the essence of the new product that it should be a mass *consumer* product, taking up slack on domestic telephone lines.

In 1974, the viewdata specification and the two standards for broadcast videotex (teletext) under development by the British Broadcasting Corporation and the Independent Broadcasting Authority were made compatible, following discussions under the auspices of the British Radio Equipment Manufacturers' Association. A two-year private trial of viewdata was conducted at Martlesham from January 1976, to gain experience of the system and to recruit IPs and television set manufacturers who would make terminals. British Telecom took a policy decision that any eventual videotex service would be offered by British Telecom as a 'common carrier'. A wider, experimental test service was to be set up in June 1978, but a few months before this began BT announced a public service to begin in 1979, irrespective of the results of the experiment.

This haste was related to a desire to establish its specification as

* British Telecom did not in fact come into being until 1981: until that date it had been under the British Post Office. However, to avoid confusion we have throughout this chapter used 'British Telecom' or BT.

the *de facto* videotex standard. In practice, the start of the test service slipped back to September 1978, and was further delayed by shortages of terminals. Shortly before the test service began, it was decided to supplement the panel of domestic consumers with business customers. Confusion arose because the test service ran in parallel with a public service from March 1979, and further confusion was added by the change of name from 'Viewdata' to 'Prestel', when it proved impossible to register 'Viewdata' as a trade mark.

It was, however, rapidly clear that there was no substantial mass market for Prestel among domestic consumers and by the end of 1980 only 1,000 of the 10,000 Prestel subscribers were domestic users. There were a number of obvious problems:

(i) *The high cost of terminals.* Early terminals were modified large-screen colour television sets in line with the general expectation that Prestel would be a conventional new consumer durable product which initially sold to the higher socio-economic groups who traditionally bought large-screen sets. Terminal manufacturers saw Prestel as analogous to the introduction of colour television—offering them a chance to make high profits on initial sales with prices subsequently falling as the mass market developed. The result was that a Prestel set cost approximately £1,000 ($2,120 in 1974), whereas the equivalent colour TV cost £350 ($742) and cheaper colour TVs £200 ($414).

(ii) *The high cost of the service itself.* British Telecom were anxious that from the start the costs of Prestel should cover themselves (unlike the West German service which fixed early charges on the basis of one million subscribers—a form of loss leadership). This meant a comparatively high rental charge as well as normal telephone call charges for use of the line when consulting Prestel. In addition, there were 'page charges' levied by the Information Provider. Given the complex indexing system and the time therefore taken to consult, say, a consumer report on refrigerators, the overall cost could easily amount to £1 per consultation (over $2 in 1979–80).

(iii) *The limited usefulness of the information provided.* Most of the information provided was reference material, relatively readily available in public libraries—for example, consumer reports, government statistics, timetables. The early service did not provide 'gateway' access to, for example, bank computers, which would enable users to check the state of their accounts, etc. Such gateway access was central to the West German Bundespost's view of its own role as a 'common carrier', and it commissioned additional software to implement gateways when it took a licence for Prestel. This has now, belatedly, been introduced into the British Telecom service.

One service, however, did flourish. This was the travel agents' on-line consultation of timetables and booking of tickets. (By 1984, 90 per cent of British travel agents used the service). As a result, business customers outnumbered domestic customers early on by nine to one and led British Telecom to radically change its strategy. In the words of the revised marketing plan produced in 1981:

> Prestel is not an integrated information service. It embraces hundreds of different services, most of which are only of use to a minority of customers . . . . We believe that each customer we acquire in the next two years will have a specific reason for obtaining Prestel . . . . Our prime task, therefore, is to identify, develop and market intensively the small number of information services which can actually clinch sales. These are the real product we are selling. [Levis, 1980]

This involved the rejection of the original, technocratic conception of Prestel as a computer data base to which domestic consumers would effectively address questions. The domestically-orientated IPs began to withdraw. In line with the new policy, fourteen computer centres were closed in September 1981 and staff redeployed within British Telecom.

Meanwhile, British Telecom introduced a cheap adaptor to television sets, regretting its earlier decision to back the manufacturers rather than its own inclination that a cheap adaptor would help develop the domestic market.[3] British Telecom executives responsible for Prestel were inclined to blame the terminal manufacturers' insistence on the large integrated colour TV/terminal for Prestel's failure to expand. As suggested above, there were other marketing failures. Nevertheless, the cheap adaptor, when finally introduced alongside (and compatible with) the very successful BBC microcomputer, helped to improve the popularity of domestic Prestel.

By 1984, the number of subscribers had increased to 44,000 (still well below the 100,000 originally envisaged within the first three years of service), with new subscribers joining at the rate of 1,500 a month (Cane, 1984). Some 63 per cent of terminals were in businesses (26,500) and 37 per cent in homes (but this sector is now expanding fastest). Charges are based on a rental charge—(£16.50 ($22)—per quarter for a business; £5 ($6) per quarter for homes in 1984), a time charge based on time of day (with no time charge in offpeak periods—i.e., again favouring the home user); a page charge set by the information provider (the majority of pages carry no cost) and, of course, the normal telephone cost for making local calls.

In addition to its 'common carrier' information services, Prestel now offers a whole range of specialist services (de Zoete and Bevan, 1984, pp. 206-7). For business, these services include:

(i) *Prestel Travel Service*, including Skyback—an on-line airline reservation service using a gateway connection to the airlines' reservation booking system.
(ii) *Prestel Citiservice*—a rapidly updated information service on stocks, commodities and foreign exchange, mostly for users outside the City of London. It is a less sophisticated version of the Reuters or Telerate services, but also much cheaper, and is mainly used by small businesses.
(iii) *Prestel Racing Service*—a specialist service for jockeys, owners, trainers and punters by Bloodstock and Racing Data Ltd. It was launched in June 1983 and by 1984 had 150 subscribers.
(iv) *Private Prestel*—for companies wishing to use Prestel for internal communication (e.g. between home and office, or local branch and central office). Again, the gateway facility enables private clients to access their firm's computer. So far there are relatively few private clients, but numbers are increasing as the service is publicised.

The residential services offered by Prestel now include:

(i) *Micronet 800*—a service launched early in 1983 providing downloadable software for home computers. This has proved a major attraction to home computer users and is one of the main factors currently boosting Prestel's new subscriber rate.
(ii) *Homelink*—a home banking service launched in September 1983. It links the Nottingham Building Society and the Royal Bank of Scotland and offers subscribers the chance of switching deposits from the (higher earning) building society account to a chequebook account with the bank. Its aim is 100,000 subscribers by 1986.
(iii) *Club 403*—this is a pilot scheme to develop Prestel as a local community link service, being run in the West Midlands (on page 403 of Prestel). It is sponsored by the Department of Trade and Industry but so far take-up has been disappointing.
(iv) *Prestel Mailbox*—an electronic mail service which allows business or residential users with numeric and alphanumeric keypads to send messages to each other and via telex.

As far as the United Kingdom is concerned, therefore, Prestel has got off to a somewhat shaky start, but is now moving ahead more strongly and beginning to develop markets in terms of specialist services which offer considerable potential for growth. Much that has gone wrong could, with forethought, have been avoided. Management initially geared itself too closely to two objectives: first, with videotex as a means of taking up the slack on domestic lines and, second, with the need to establish the service quickly in order to dominate international standards. Looking at it in terms of the *filière*, the upstream technological elements were allowed to dominate the downstream marketing element. It is an interesting commentary on the relative stage of product maturity (or perhaps on the relative insignificance of the videotex *microfilière per se*) that this proved so disastrous a product recipe. The result effectively meant a false start and the loss of two vital years in building up a clientel for a service that is likely to be rapidly overtaken by broadband cable and satellite networks.

WEST GERMANY

The West German videotex project, called Bildschirmtext (BTX), was only introduced as a nationwide commercial service in June 1984. It is therefore too early to tell how successful it will be. This short description will therefore concentrate on the history of the development.

The Deutches Bundespost (BDP) was immediately interested in the videotex system once British Telecom announced its service publicly in 1976. In January 1977 it bought the Prestel software, together with a GEC 4080 computer, and in September 1977 had the first public demonstration of the system at the International Radio and Television Exhibition in Berlin. There was considerable enthusiasm from potential information providers, but another problem immediately arose which was to bug the project for the next six years—the question of control. The *Länder* (regional goverments) claimed that videotex should be, like the press, subject to regulations which they laid down. The DBP claimed that, like postal mail, it was a private exchange of information, with the DBP solely responsible. It was not until March 1983 that agreement was finally reached defining the status of BTX and with the *Länder* abandoning most regulatory powers.

The DBP strategy in relation to videotex was twofold:

(i) In order to overcome the *Länder* objections, it was important that videotex was introduced as a national service from the start (even though the Prestel architecture made it more suitable for introduction as a regional service). Hence the importance of the 'gateways' which gave access to national data collections. The DBP therefore immediately commissioned the design and development of gateways which were, ultimately, to prove so valuable to BT's own marketing programme. Ironically, they were commissioned to meet regulatory, not marketing, objectives.

(ii) The DBP was also anxious to get a test project off the ground and this was introduced in June 1980 in Düsseldorf and Berlin, the former because it represented a good average mix of the West German population; the latter because its population was thought to be more receptive than most to innovation. The trial was originally due to begin in January 1980, but had to be postponed for six months because of regulatory problems and the failure of the TV (terminal) manufacturers to supply sets adapted as required.

Technically, BTX was a success. The 'gateways' were designed by a subsidiary of GEC[4] and the Verbrauchers Bank was the first to use the facility (October 1980), though rapidly followed by the other big banks (Dresdner, Kommerzbank, etc.) By 1984, more than sixty gateways were available. The DBP also went to some trouble to make sure it had an appropriate network architecture. Once the service was to be available country-wide, it was important to have good connections between local centres. The DBP went to public tender and the contract was won in 1981 by IBM-Deutschland with a design based on the assumption that 98 per cent of usage would be at local level (handled by minicomputers), while a large IBM 4300 based at Ulm would provide the central data base which could be accessed by local minicomputers. The IBM solution was more flexible and cheaper than the original GEC version and could also operate through the packet switching network (which further reduced long-distance costs). But the DBP saw another advantage in using IBM: 70 per cent of computers in West Germany are IBM computers, hence interface would be easier. The choice of IBM by the DBP came as a surprise: until recently the DBP has always automatically bought German and has had a common research link on videotex with SEL, the German ITT subsidiary (see Chapter 4). Rivals argued that IBM had gained the contract by offering a 'dumping' price and promising schedules which could not be met, a criticism which gained further credence when the introduction of the service had to be delayed from September 1983 to June 1984 *because* the IBM network was not ready. IBM incurred a DM3.6m penalty to the DBP as a result.

The DBP also tackled the issue of the European standard. With two rival standards—Prestel and the French Antiope—discussions within CEPT (Conférence Européenne des Postes et Télécommunications) had been protracted. A compromise was finally reached in May 1981, based on a German proposal which is compatible with the two systems, but leaves the door open to further development. (In fact, this norm has remained largely theoretical since the only PTT to have committed itself to it is the DBP!)[5] The CEPT norm is basically incompatible with the Canadian norm of Telidon, the major competitor to the European systems.

The other issue to be tackled was tariffs. During the trial, costs to subscribers had been very low, with a monthly fee of DM8 ($3.3) and 90 per cent of the information provided was free. Those who had participated in the trial had been provided with a built-in decoder to their TV free of charge[6] and 40 per cent of these declared that they would be ready to pay more for BTX services with an upper limit of DM35 ($14.4) per month (Forschungsgruppe Kammerer, 1983). For the IPs, the sunk cost of building up the access system was between DM5,000 and DM1m ($2,000–$400,000), the latter being linked to use of a gateway, but for the user (during the test) the first forty hours of usage were free.

As noted, tariffs were one of the main reasons for Prestel's initial failure. The DBP avoided the same trap as Prestel by setting its tariffs at a level designed to recoup costs, with a million subscribers, with further 'loss leader' incentives (for example, electronic mailing free until 1985 and at half price from 1985 to 1986). The IPs have formed themselves into an informal grouping and it seems that pressure from this association has played a significant role in keeping tariffs low (IHK, 1983).

In spite of the eight months delay in introduction, there is considerable optimism in the DBP about the future of BTX. Forecasts speak of between 150,000 and 350,000 subscribers by the end of 1985 and between 1 million and 2.3 million in 1988 (IHK, 1983, p. 20). By August 1984 subscribers number just over 18,000. Much depends on the price of the terminal—and the service trial did not help here. To date, no major manufacturer has managed to produce a set incorporating a decoder for less than DM1,000 ($385). To overcome this, the DBP, like British Telecom, commissioned a cheap decoder chip. But again there was trouble with time. It was not available until the Autumn of 1984, but the cost was to be DM600 ($230). Siemens, meantime, has helped introduce a decoder called Mupid, developed by the Technology Institute of Graz (Austria), which is half modem and half microcomputer and capable of software retrieval. This is seen as enhancing the prospects for the synergetic development of microcomputers and videotex Personal computers are now one of the fastest-growing areas of expenditure in West Germany.

Overall, it would appear that the DBP has learned the lessons of British Telecom's mistakes with Prestel. They have given care and forethought to network architecture and to testing consumer and IP reaction. Nevertheless, given the not very enthusiastic reception during the trial, when prices were lower than in reality, they should be wary of over-optimistic forecasts. The future is obviously difficult to predict.[7] On the supply side, the technological development of the telephone system and in particular the introduction of the Integrated Service Digital Network (ISDN) (see Chapter 4) widens the range of service available, while the growth of local area networks (within offices) and home computing will affect demand.

With regard to the *filière*, the development of BTX has several interesting features. Clearly, the DBP dominated the *filière*, but its relations upstream and downstream have changed over time:

(i) With the development of gateways, the role of actors close to final demand (the IPs) has increased.
(ii) In its relations with software providers, the DBP found itself caught up by the web of vertical relationships. By going out to public tender, the DBP behaved opportunistically (Williamson, 1979), particularly in relation to SEL (the ITT subsidiary in Germany) which had adapted the original GEC software for use by the DBP. Equally, IBM behaved opportunistically once it had secured the contract and found the DBP dependent on its ability to deliver.
(iii) In contrast, in its relations with TV manufacturers, the DBP has deliberately chosen not to exploit its position within the *filière*. On the contrary, its lack of leadership has left terminal manufacturers with little interest in the project and unable to meet hoped-for (but not explicit) goals within the programme.

What perhaps stands out above all from this experiment is the difficulty a public enterprise such as the DBP faces in this situation. Its ability to implement a successful videotex programme is limited by its inability to integrate upstream or downstream within the *filière*. It therefore becomes vulnerable to the opportunism of other operators, while at the same time its very nature, as a public body with (implicitly) the public purse available to back its ideas, biases the whole experiment. The fact that the DBP were going ahead with

BTX was never really in doubt, as evidenced by the DM500m ($200m) it will have sunk into the project by 1985. The test programme was not a market simulation (to find out whether or not there was a market for the product) but an attempt to find the quickest and easiest way to implement the project. In this sense the evolution of the whole *filière* has been biased by the presence of the public monopoly: equally, given its position within the *filière*, the public monopoly (in this case the DBP) might well have found it more profitable to limit its functions to that of common carrier—its basic role anyway within the telephone network.

FRANCE

The French videotex system—called Antiope-Télétel—is the second major system in Europe after Prestel and the latter's main rival. Its commercial introduction lags behind Prestel and its sales to the other European PTTs have been much less successful—only the Spanish PTT having bought it. From a technical point of view, however, it has features which are superior to those of Prestel—for example, by using the packet switching network, Transpac, its tariffs are lower and it offers universal access to external computers and data bases without having to worry about gateway facilities (which are expensive to install). It is also, uniquely, linked to the PTT data base for the whole telephone system and therefore offers the facility of the electronic directory which in turn has far-reaching consequences for diffusion.

Antiope-Télétel developed from two different sources. First, in 1971-2 came experiments by CNET (the public telecommunications laboratory in France) aimed at developing a cheap terminal for time-sharing use using telephone, television, modem and keypad. Second, came experiments from the CCETT (Centre Commun d'Études de Télécommunications et de Télévision) for a broadcast videotext service called Télépresse. British Telecom's announcement in January 1976 that it was going to introduce a videotex service based on Sam Fedida's ideas brought a merger of the CNET and CCETT initiatives in order to make broadcast and interactive videotext compatible under the generic name Antiope. Broadcast Antiope was presented in Moscow in September 1976; interactive Antiope appeared for the first time at the Berlin Radio and Television Exhibition in September 1977.

After technical implementation came commercial application. The 'Plan Télématique' announced in March 1978 (see Chapter 4—

pages 115–119) announced two important initiatives on interactive Antiope:

(i) A commercial service called Télétel was to be launched on a test basis in Velizy. Versailles and Villacoublay—the aim being to involve 2,500 volunteers (this was called the Télétel 3V trial). The terminal designed for Télétel 3V was an adaptor to be plugged into a TV set.

(ii) A trial was announced for St Malo and (later) the whole district of Île et Villaine, involving 250,000 subscribers for an electronic telephone directory. Subscribers were to be provided with a small, low-cost, terminal free of charge called Minitel, through which they would be able to access any telephone number in France for the cost of a local call. Minitel was to be compulsory and would, it was hoped, rapidly replace the paper directories.

The two projects were closely linked. The software was the same and the aim of the compulsory electronic directory was both to stimulate the production of cheap terminals and to accustom people to using them. Ultimately, it was assumed that some 30 million Minitel terminals would be required and it was hoped that each would cost no more than 500 frs ($110). (Minitel was not, of course, a TV set but a small desk-top VDU with a full keyboard, offering only black and white, however, and being much smaller, poorer quality resolution than a TV set. For most videotex purposes, for instance electronic banking or travel booking, this is sufficient).

The Plan Télématique, which was endorsed by the Cabinet in September 1979, ran into fierce opposition. On the one hand, this came from the press, which feared an erosion of advertising revenues; on the other, from the Socialists who objected to the DGT's (Direction Général des Télécommunications—the Telecom authority under the PTT) steamrolling the ideas through without proper discussion or consultation. The Government had to agree to a strict evaluation of the Velizy and St Malo trials and the technical separation of the two projects, with only the terminals (not the software) being compatible.

The election of May 1981 and the new Socialist Government brought further changes. The electronic directory was to be voluntary, not compulsory. This of course meant far fewer were included in the trial. By mid-1983, only 14,000 subscribers had been provided with Minitel. Meanwhile, technical delays meant both trials fell behind schedule. The Télétel 3V trials did not begin until June 1981. Orders for Minitel were placed in April 1981, after a public tender which had been won by the CGE subsidiary Telic at a cost of approximately 3,300 frs ($600). In 1982 came the second round of orders, again for 300,000. This time the contract was shared between three suppliers—Telic, TRT and Matra—with the price down to 1,200 frs ($182). The next generation of Minitel to be ordered (450,000 units) in 1984 is to include some processing functions as in a small microcomputer.

Meanwhile, of course, the Télétel 3V trial was proceeding. Like the BTX trial, the sample was limited and did not increase much (2,187 in November 1981 to 2,280 in December 1982), with an average of only two calls a week. Services on offer increased from 84 to 180, and included an 'electronic newspaper' put together by 88 agencies, including press clubs, banks, travel agents, etc. Of the calls made, 30 per cent were for press information, 10 per cent for financial information and 8.5 per cent for the 'electronic pigeon hole' personal message service—all roughly comparable with the BTX and Prestel experience. As in these two cases, there were problems with the terminals, it being clear that the TV set decoder was much too expensive *vis-à-vis* Minitel. Thomson and TRT had to abandon the production of these decoders. But on the whole the experiments were encouraging and further trials of 'convivial telematics' were launched in Grenoble and Nantes in 1982.

In relation to network planning, two major decisions were taken. For Télétel there was to be no public access (gateways) between data bases and end-users, but access was to be via the Transpac system, with tariffs dependent on the volume of data exchanged (not, as with the telephone, on time and distance). This provides a far more flexible system than the Prestel plus gateway system developed for BTX and allows a much easier interface between private videotext systems and the public network. In November 1982, using this system, the DGT opened a new service of 'professional videotext' which consists of renting a Minitel terminal for 70 frs ($10.65) a month in areas where the electronic directory is not available, and allowing business users to build up their own information networks via Transpac. The service was extended to the whole of France in January 1983 and, according to *Le Monde*, by September 1983 130 private videotext systems operated via this route.

The other major decision related to the electronic directory where the same (Transpac) system enabled access to the national subscriber data base, updated every five days. In 1987, the electronic directory will be available everywhere in France, with 1.5 million subscribers having a Minitel terminal by the end of 1985.[8]

It was recognised by 1983 that, whatever the Government might have said in 1980, Télétel and Minitel were part and parcel of the same thing and that the two programmes should effectively be merged. This was acknowledged recently when the bank, Crédit Commercial de France, told its customers who owned a Minitel terminal that they would be able to use it for making transactions. Another development is the 'smart card'—an enhanced credit card with a microprocessor and memory embedded into it. Coupled with a Minitel terminal, it will allow booking and payments to be made from home.

The cost of the videotex developments in France has been huge. Glücksmann (1982) gives figures of 60m frs ($14.1m) for 1979, 210m frs ($49.7m) for 1980, and 230m frs ($42.4m) for 1981. Other commentators have suggested that by 1982 Télétal had already cost 1.6bn frs ($250m), while the overall telematics budget for 1983 was 1.5bn frs ($197m at 1983 exchange rates) and 2bn frs ($263m) for 1984.[9]

Overall, its two innovations give the French system considerable advantages over its competitors. The 'common carrier' solution via Transpac makes the French system closer to its American counterparts, which offer special services to limited numbers of subscribers, but because it operates via its packet switching network, the French tariffs can be lower. The electronic directory enables the DGT to dominate the *filière* to an extent that neither British Telecom nor the DBP found possible—the DGT were able to specify precisely the type of terminal required and to update specifications with each new round of ordering. The same is true of its upstream linkages. It is able to specify both computer hardware (from CII) and software, and since the French CII-Honeywell Bull depends on the French Government for its future order book, it is not able to behave opportunistically as IBM did with the DBP.

Of all the innovations, the electronic directory is both the boldest and likely to have the biggest pay-off. It has the great advantage of opening up the market for other services (such as banking or travel) and thus enabling other services to benefit from the externality; once a customer is given a terminal and gets used to handling it, it is much easier to provide additional services. Compare this with the development of Prestel in its first two years where there was the wish to develop the mass market but no 'loss leader' services encouraging people to experiment.

Nevertheless, from a pure *filière* point of view, it is not clear that the DGT *should* have such control downstream—rather, given the uncertainty of outcome, one might have expected a strategy aimed at maintaining maximum flexibility (as in the United States). Once again, it is the fact that the DGT is a public monopoly that is the decisive feature. The initial project, with the compulsory Minitel terminal, presented almost a caricature of the public monopoly—able to enforce a particular mode of behaviour on the citizen. Cherki (1982) argues that the DGT planned the electronic directory in exactly the same way as it planned its investment programme for the telephone network five years before. Unlike the telephone, however, there was no pre-existing demand and because of its nature (linked to mass media) it sparked off strong social reactions. The industrial goal sought by the DGT—the creation of a strong terminal industry—was not, according to Cherki, worth the social friction it created. In this sense the episode represents very much the 'apprenticeship' of the DGT in its relations with French society.

From an economic point of view, it is also hard to justify the DGT's strategy. To think in terms of providing 30 million Minitel terminals by 1990 is highly speculative; even the scaled-down trial of Île et Villaine has shown up technical problems. (It is, for example, clear that the idea of a 'trial' with 250,000 subscribers would have been impossible to handle and would instead have delayed the nationwide introduction of the service.) Indeed, the whole project would have involved very considerable expense at a time when the DGT was seeing its profits eroded for other reasons (see Chapter 4). The manufacturing of cheap terminals, although important for videotex, was by no means the most important aspect of the telematic challenge, and in any case was arguably not achievable: the price of Minitel is still three times higher than initially planned.

With the redefinition of policy after May 1981, the pace of implementation is now in tune with the DGT's technical and economic capabilities. The Île et Villaine test has shown that, given the average frequency of calls experienced there, the Télétel project is just breaking even (Leclercq 1983). Any acceleration would increase the costs of implementation.

As for the other actors in the *filière*, the IPs were present only in the Télétel 3V trials when 180 of them tested their services. As in Germany and the United Kingdom, they seem to have found difficulty defining a product to sell and complain in their turn of the difficulty of coping with the medium. It would appear that the relative passivity of the IPs is the counterpart to the entrepreneurship of the DGT. Whether this is a good thing has yet to be seen: once the Minitel project makes the electronic directory available throughout France, it will need other services to turn it from a gadget into a consumer durable. But the Minitel terminal still lacks some essential features, for example, storage capabilities for information, which will be required if this transformation is to succeed.

To sum up, French policy towards videotex has taken maximum advantage of the public monopoly status of the PTT. The early project, with its compulsory Minitel electronic directory, rapidly proved to be a social failure (and would probably have been a major economic one too). The success of the amended programme now depends on a switch in leadership—from the DGT to the IPs. The latter have to convince the public that videotex can provide them with a range of services which they cannot get (as conveniently) from any other source. The uniqueness of the French experience lies in the fact that all the actors in the *filière* can be fairly confident that in three to four years there will be a large market for videotex services via Minitel, combined with considerable flexibility on the supply side. The great advantage of the electronic directory is that, although in its revised form it does not remove uncertainty, it has nevertheless considerably decreased uncertainty within the transitory phase of development.

Lessons and conclusions

The lessons to be learned from the videotex experience are lessons about product innovation by a public monopoly. In a competitive or oligopolistic market, consumer signals, in the form of purchases or refusals to purchase a new product, guide suppliers and new product offerings are adjusted accordingly. By its very nature, monopoly crushes signals from the market. This is not simply a matter of statutory monopoly: it is endemic to systems such as telecommunications networks that the scope they provide for experimentation with alternative products is limited. It is difficult, for example, to imagine a PTT setting out to offer alternative videotex systems in the hope of finding a product recipe acceptable to users; the best they can do is to offer a different product to different groups of users. In the event, instead of various different types of system being tried out at low volume, the PTTs tried to leap-frog the product experimentation stage and to impose, from the start, their own preferred product recipe. That they ran into difficulties and had to backtrack and redefine strategies is hardly surprising.

In this respect, it might be worth briefly contrasting European experience with that in the United States and Canada, where the telephone authorities have taken a much less prominent role.

In the United States, videotex does not exist on the same scale as in Europe.[10] For many years there have been some videotex services such as services which use an ASCII keyboard to send messages via telephone and TV. One of the largest of these—the Source (32,000 subscribers)—is run by Readers Digest and allows users of personal computers to access data bases and a personal message service via Tymnet and Telenet. The largest—Dow Jones—is a specialised service sending stock market news to about 42,000 subscribers with personal computers or access to a computer terminal. Interactive videotex services are a relatively new phenomenon and are being promoted by a variety of companies. Viewtron in Florida has 700 subscribers and uses AT&T terminals; AT&T and CBS have joined forces at Ridgewood, New Jersey, with a trial for 200 households. AT&T has chosen a standard compatible with the French Antiope and the Canadian Telidon (but not with Prestel or the CEPT norm). The Times-Mirror experiment in Los Angeles is on trial in 350 households; GTE has Prestel rights and even IBM recently announced that it was moving into the field. But all these experiments are, to date, small and diffused, and many of the difficulties they face such as the price of terminals and cost-overruns are reminiscent of the European experience.

In Canada, the most important feature of developments is the unifying influence of the Telidon system, developed by the Department of Communications at a reputed cost of Canadian $9m. The cost of linking all the local videotex systems developed by regional concession holders has been approximately $ Canadian 1bn (Ratzke, 1982). Telidon provides the norm for all these systems (i.e. they have to be compatible with it), but as a technical standard it has a number of advantages, not least a better visual display,[11] and as a system it is on trial in a number of provinces. Bell Canada has also launched a new system—VISTA in Ontario and Quebec. Several provinces— Ontario, Manitoba and British Columbia are examples—currently support experiments run by the different 'common carrier' telephone networks. Among these is the famous 'grassroots' service in Manitoba with 150 terminals linking scattered agricultural communities and providing local news and advice sessions.

The essence of the Canadian approach is unity in diversity. The high technical quality of Telidon provides the overall framework, but it is one which allows for diversification both of services offered and transmission facilities: it is compatible with cable, telephone circuits, leased lines or fibre optics. It is also selling well overseas; the Times/Mirror experiment in Los Angeles uses Telidon software, while in Japan both Mitsui and Mitsubishi are developing Telidon for the Japanese market.

What is interesting in the comparison of North American with European experience is the difficulty all the promoters seem to be experiencing in getting the service 'off the ground'. It is probably fair to say that the PTTs and the computer manufacturers were the only actors capable of producing the videotex technology. The computer manufacturers had little motive for doing so. Videotex involves a cheaper and simpler version of their existing terminal-based products, and the total market it would create for computers would not be a sufficient fraction of their output to merit experiment. This lack of interest on their part is reflected in their distance along the *microfilière* from the end-users (see Figure 4.2). By contrast, the PTTs were able to dominate the *microfilière*, and it is this domination that has shaped their product strategy.[12]

The whole point of the discussion, once the PTT domination of the *microfilière* is recognised, is how the vertical links shown in Figure 5.2 were handled. In the British case, the integration (or better, quasi-integration) backwards with GEC worked reasonably smoothly, while the forward link with the IPs and terminal suppliers did not work well. The former were too closely tied to British Telecom's concept of Prestel as a mass information provider, and it was not until BT changed its mind that greater flexibility began to emerge. As for the terminal suppliers, they were neither integrated into the Prestel project nor strong enough, in either a technical or market sense, to ignore what BT was doing. Had they identified more closely with BT's initiative and put strong emphasis on the technical features—above all the design of a cheap decoder—then their market position *vis-à-vis* both European and Japanese competitors would have been much improved. Such a strategy would have required co-operation with one of the British semi-conductor companies (e.g. GEC Semiconductors). In the event, given the passive role adopted by the terminal manufacturers, the initiative had to come from BT.

In the German case, the backwards integration with IBM did not work well, but the forward integration with the IPs was better, thanks to the introduction of the gateways. The links with terminal suppliers, although closer than in the British case, were nevertheless not close enough to avoid the bottleneck of terminal costs. With its bold investment in BTX, the DBP reduced the market uncertainties for manufacturers, but not enough to eliminate the direct link with end-users and hence doubts about the market would materialise.

By contrast, in the French case there was maximum integration both backwards (to CII-Honeywell Bull) and forwards to the terminal suppliers (Matra, Telic and TRT), who in effect became subcontractors to the DGT. Flexibility was guaranteed by the introduction of access via Transpac (packet switching system) to large data collections held in computers. But the DGT has not managed its own Télétel/electronic directory project well, mainly because from the outset it treated the two as distinct systems instead of one integrated system. As a result, IPs focused on Télétel, and failed to recognise the potential that the electronic directory, Minitel, offered for attractive and cheap services.

In the United States and Canada, the 'common carrier' network services do not dominate the *microfilière* in the same way. In the United States, it is the IPs who seem in many cases to be the 'leaders', which in turn narrows the scope of the service (hence the number of special user group projects), but ensures a service better tailored to users' needs. Integration is often achieved via operation on a joint venture basis (e.g. CBS and AT&T). The Canadian structure is half-way between the European and American models. It gives the opportunity for vertical integration with the common carrier, but does not require it. Centralisation via the technical norms of Telidon is set high in the *filière* and allows for flexibility both in network and service provision. Indeed, in some senses, it would appear to combine the advantages of both the American and European approaches. Outwardly, no control at all seems to exist within the whole *filière*, but the unifying concept of Telidon makes the market much less risky for terminal suppliers.

What really distinguishes videotex in Europe from its North American counterpart is therefore the role of the PTTs. It was the PTTs' conception of videotex as a mass market product that dominated the 'product recipe' offered and because of their public service monopoly position they were able in effect to impose this product recipe on consumers.

There are two interesting aspects to this situation. In the first place, given the dominant position of the PTT in the *microfilière*, it seems unlikely that any private initiative could have got off the ground without the active participation of the PTT. Even if the PTTs had limited their involvement to that of common carrier (in effect what the Bundespost is trying to achieve), their intermediary role between IPs, customers and terminal manufacturers would inevitably have brought them into prominence. Second, it is worth pondering the motivation that led the PTTs to take the initiative, and here the answer, at least in the case of the United Kingdom and France, would seem to be industrial policy. British Telecom saw its active stance on videotex as not only a useful method of using the spare capacity on domestic lines, but as simultaneously promoting the interests of British terminal manufacturers by providing them with the 'next generation' (after colour TV) of TV sets. By pushing ahead fast and establishing Prestel as the *de facto* international standard, they offered the same manufacturers a chance of penetrating world markets. The French were more explicit in their aims. The rapid development of Antiope was expressly to prevent Prestel from becoming the world standard: similarly, the Minitel experiment deliberately aimed at leap-frogging French terminal manufacturers into the forefront of terminal (and VDU) suppliers. The German situation was different. The Bundespost did not have the industrial policy aspirations, but once 'bitten' by the idea of BTX, was loath to let it go, doggedly pushing ahead in spite of the difficulties. Underlying such behaviour is, perhaps, the technical bias of the engineers who ran the PTTs. Videotex attracted them because it was technically simple and therefore ought, in their eyes, to be a success. What they were not aware of, because they were not experienced in such matters, were the problems of *marketing* their new playthings.

Finally, it is perhaps worth pondering the transitory nature of such developments. A decade ago, when videotex was in its infancy, it seemed promising because the costs of recabling seemed too great to bear. Now, recabling is a live issue in many countries, opening up the possibility of high-bandwidth services which bypass the very bottleneck that videotex was intended to squeeze through. It is easy to be wise with hindsight but readers in the future may be forgiven for wondering what all the fuss was about.

Notes

1. In Finland, in fact, the videotex service is run on a 50-50 basis by the PTT and the major press group Sanoma.
2. One of the criteria against which patents are judged is 'obviousness'. Anything seen to be obvious is not patentable.
3. The hope was that market penetration could be achieved through the price advantage of adaptors made in much larger volumes than integrated terminals, and in 1979 Telecom asked for tenders for an option on 200,000 adaptors to cost up to £50 ($106) each in that volume. Ayr Videotex apparently specified such an adaptor which was acceptable, but in February 1980 British Telecom decided not to take up this option following pressure from the television set makers (*New Scientist*, 30 October 1980, p. 302).
4. The software for these gateways was sold by the DBP to British Telecom in 1981, but without patenting it—an indication of the poor marketing capability of the DBP. Like many PTTs, the DBP suffered from having only sold the telephone service, where demand was ready-made. As is made clear

in Chapter 4 on Telecommunications, this experience often left them at a disadvantage when it came to handling big business (e.g. IBM) or attempting to diversify their product range.
5. The decision to go on to the CEPT standard brought difficulties for the IPs and gateway users. They had developed their software for the original GEC/Prestel alphamosaic standard, and then had to move to the CEPT/SEL/GEC solution in the transition period before opening, then had to redesign the whole package for the CEPT/IBM standard which, from the end of 1984 was the only one available nationwide (*Wirtschaftswoche*, 10 August 1984).
6. A market experiment began in Düsseldorf in June 1980. Participants in the trial paid the same price for their terminal as for a comparable colour television, with the Bundespost paying the balance of the terminal's actual cost direct to manufacturers. It proved difficult to recruit a sample, and eight months into the experiment only half the target number of participants had been found. The reluctance of consumers to participate was reinforced by the non-commitment of potential terminal-makers such as Grundig and Blaupunkt who were sceptical about the potential of the system. Moreover, there was no incentive for retailers to try to sell the system, since their mark-up was exactly the same as for colour TV sets.
7. The most recent forecasts still put subscribers at 300,000 by 1985, but suggest that between 70 and 90 per cent of these will be business users, while the 1981 forecast put households at 60 per cent.
8. Events in June 1985 illustrate that perhaps here too the DGT's plans were too bold. The Transpac network has been unable to cope with the traffic created by the Minitel electronic directory terminals, and waiting time for access has lengthened considerably, leading the DGT to announce the suspension of all new Minitel terminal deliveries until September 1985. The reasons for the failure are complex, and not all to be found in the DGT's overambition. Transpac is a network basically designed for data transmission by business users. (It is in fact the largest packet switched network in the world, four times larger than its nearest rival, the US Tymnet.) It is not therefore wholly suitable for the sort of demands being generated by residential customers. In other respects, Minitel is the 'victim of its own success' (in contrast to Prestel's slow take-off!) and one of the paradoxes of this incident is that it gives gratuitous publicity to the programme.
9. See *Le Monde*, 18 August 1983 and 22 September 1983. The figures include the videotex access points created throughout France in order to allow access to the service on a local call basis, plus the cost of the free Minitels.
10. It is sometimes argued that European countries have taken the lead in videotex because American companies do not believe in its success. This is wrong: private initiatives are numerous in the United States, and AT&T—which could have developed a videotex product—was prevented by regulatory legislation from offering such a product.
11. Telidon uses a display technique known as alphageometric, which uses geometric elements (joints, corners, lines, curves) as elementary units in compound signs, letters and graphs. It gives much better resolution than the

European alphamosaic system, but uses much more memory (24K, against 2K for Antiope and 1K for Prestel) and transmission time is slower.
12. IBM's decision of the spring in 1985 to team with Merrill Lynch in the provision of on-line financial information services (which are in many respects similar to videotex) confirms this. The success of the IBM PC in the United States, particularly in the business community, makes worthwhile this extension of IBM interests. But it happened *after* the diffusion of PCs, and did not come as a part nor as a condition of this diffusion. And IBM is now trying to enter the *microfilière* from its strong position as terminal provider *not* as computer manufacturers—a clear sign that the control of the *filière* is located downwards, close to the end-user.

Bibliography

Cane, Alan (1984), 'Videotex family puts down its roots', *The Financial Times*, 25 July.
Cherki, E. (1982), 'Le project télématique: réflexions sur une stratégie industrielle', *Bulletin de l'IDATE* No. 9, October.
Fedida, S. and Malik, R. (1979), *The Videotex Revolution*, London, Associated Business Press.
Forschungsgruppe Kammerer (1983), *Wissenschaftliche Begleituntersuchung zur Erprobung von Bildschirmtext in Berlin*, Munich.
Glücksmann, R. (1982), *Telematica: dal Videotex all' Office Automation*, Milan, Gruppo Editoriale Jackson.
IHK (Industrie in Handelskammer) (1983), *Bildschirmtext: Erfahrungen und Marktchancen*, Berlin.
Leclercq, D. (1983), 'Une utilisation à grande échelle du videotex: l'annuaire électronique', *Informatique et Gestion*, December.
Levis, Kieran (1980), 'The Marketing of Prestel, 1980–81', mimeo, London, the Prestel Organisation.
Malsot, J. (1980), 'Filières et pouvoirs de domination dans le système productif', in 'Les filières industrielles', *Annales des Mines*, January.
Mistral, J. (1980), 'Filières et competivité: enjeux de politique industrielle', in 'Les filières industrielles', *Annales des Mines*, January.
Post Office Research Review (1978) 6.
Ratzke, D. (1982), *Handbuch der Neuen Medien*, Stuttgart, Deutsche Verlags Anstalt.
Soulié, D. (1983), 'Comportement des entreprises le long des filières de production', Working Paper, Groupe de Recherches on Sciences Sociales, Université de Paris-Dauphine, Paris.
Toledano, J. (1978), 'A propos du concept de filière', *Revue d'Économie Industrielle*, December, pp. 149–58.
Williamson, O. E. (1979), 'Transaction Costs Economics: The Governance of Contractual Relations', *Journal of Law and Economics*, pp. 233–61.
de Zoete and Bevan (1984), 'British Telecom', report by J. Summerscale and C. Wells, published by London brokers de Zoete and Bevan.

6 Biotechnology: watching and waiting
Margaret Sharp

Biotechnology is not in a strict sense a new industrial activity. In the first place, it is not an activity but a technology, the application of which gives rise to activity in many industrial sectors. In the second place, it is not new. By its broadest definition—the technological application of biological science—it embraces agriculture, forestry, and techniques such as selective breeding and fermentation which the world has practised since pre-history. Today, over 40 per cent of manufacturing output is biological in origin. What is new is the world's increasing understanding of biological phenomena and in particular of the complex functioning and chemistry of the cell. As a result, the human race is now in a position both better to exploit the natural processes of biology and, most significantly, to manipulate their functioning. The result is a new and widespread interest in the potential use and application to industry and agriculture of biological processes, an interest which in turn is leading to increasing sophistication in the use of these processes.

While these developments are already having a marked effect upon some old industries (for example, tissue culture techniques are transforming the horticultural industry), their impact upon other industries (such as the chemical industry) is still in its infancy. One of the most important breakthroughs in technique, the Cohen/Boyer use of restriction enzymes as a route to genetic engineering, dates only from 1973 and is therefore, even as a *laboratory* technique, only just over a decade old. Many of the new industrial activities that we shall be looking at in this chapter are, therefore, also new scientific activities. Herein, however, lies their interest, for a study of biotechnology in its present phase encapsulates many of the issues which arise when a sudden spurt in scientific/technological development opens up new potential across a wide span of industrial activity —issues of the relationship between science and technology: the uncertainties of scaling-up techniques developed in the laboratory, the problems of finance, the role of goverment and research institute, and the relationship between industry and academia.

The first part of this chapter gives a brief survey of the scientific developments which have led to this new interest in biotechnology and the second considers their impact upon different industrial sectors. The third section focuses on developments in Europe and considers the strategies adopted by firms and governments. The fourth considers the role of the EEC and the case for a European initiative in biotechnology, while the final section takes a broader look at the role of governments in promoting a new technology and the general position of Europe *vis-à-vis* the United States and Japan.

What is biotechnology?

Biotechnology has been referred to as a 'jelly-industry', which is to say that it spreads itself over many activities and defies precise definition of content or practice. This is not for lack of trying: more or less every report on the subject begins with an attempt at definition. This chapter uses a fairly broad definition of biotechnology derived from the Spinks Report (ACARD, 1980): 'the application of biological organisms, systems or processes to manufacturing or service industries'. This definition reflects the fundamental unity in the origins of living materials, with advances in one sector often having implications in others. It reflects, too, the concern of biotechnology with biological organisms, such as bacteria, yeasts and fungi, and products derived from these organisms. It also indicates concern with industrial, not just laboratory, application. Microorganisms have been being used in fermentation since time immemorial: indeed this use constitutes what has been called the *first generation of biotechnology* which lasted from pre-history until the 1940s. The discovery of penicillin and the intense interest that this aroused in microbiology and the potential use of natural microbes as therapeutic or chemical agents heralded the *second generation of biotechnology*, which lasted until the 1970s.

The *third generation*, which is sometimes called 'the new biotechnology', is the subject of this study. It dates from the introduction of genetic engineering in the early 1970s. It is wrong to see this as a complete break with the past: rather, it is a coming together of first and second generation biotechnology with the new techniques of genetic engineering.

The new biotechnology in fact builds on developments in five different areas (Cantley, 1983):

(i) fundamental progress in understanding the chemistry and functioning of the cell, and particularly in understanding

the role of DNA as the molecular carrier of stored genetic information;
(ii) techniques based on microbiology for the screening, selection and cultivation of useful cells or micro-organisms;
(iii) techniques for the manipulation, alteration and synthesis of the genetic material in cells in such a way that the functioning of the cell is modified (e.g. so that the cell 'manufactures' a different product, or produces more of the same product);
(iv) techniques for plant cell and tissue culture and their application to crop cultivation;
(v) development of chemical engineering techniques both in fermentation technology *per se* and in the downstream processing for extraction treatment and purification of useful materials.

Techniques (i), (ii) and (iv) in many respects represent the constant, incremental pushing forward of the frontiers of science. Better equipment and computers have provided the tools for the constant probing, experiment and testing which has gradually extended knowledge both of the extent and variety of living organisms that inhabit this world, and of the construction and functioning of these organisms, focusing in particular on the functioning and chemistry of the cell. The net result has been to revolutionise understanding both of the workings of the human body and of the environment we inhabit.

Similarly, with technique (v), progress in chemical engineering also represents a continuum of incremental change. The interest in enzymes aroused in the 1950s by developments in microbiology led to improved methods of enzyme production and use, including the development of immobilisation techniques which enabled the enzyme catalysts to be re-used (rather than, as with rennet in cheese, to be lost in the subsequent product). Immobilisation led to the development of a new generation of continuous fermenters, now called bioreactors. The use of new biological sensors, combined with the development of microelectronic controls which can provide continuous monitoring of the fermentation process, has led to the introduction today of yet more sophisticated equipment. Meanwhile, experience in handling an increasing flow of bulk fermentation products, from biomass alcohol to single cell protein, is providing the stimulus for the improvement of the downstream end of processing—in particular, problems of coping with aqueous solutions, where extraction of the product may be both difficult and highly sensitive to handling techniques.

Technique (iii) above—methods for the manipulation, alteration and synthesis of genetic material—is the odd one out of this five and constitutes a much greater 'discontinuity' in development. This is the technique which, under its popular name of genetic engineering, has caught the public imagination. It should, however, be stressed that, important though the techniques for genetic engineering are, they could not have been developed or applied without the concomitant developments in cell chemistry, microbiology or processing technology.

There are two main approaches to genetic manipulation. One is called recombinant DNA (rDNA) and derives from the unravelling of the double helix structure of DNA and subsequent research linking the nucleic acid sequencing in DNA with the production of proteins. A particular sequence of nucleic acids within the DNA chain can be associated with the production of each particular protein. Recombinant DNA achieves the isolation of a particular sequence of DNA (for example that producing the protein insulin) and its implantation into a 'host' cell, usually a simple microorganism such as the E. coli bacterium or yeast, which is capable of multiplying itself very fast. As the host multiplies, so it produces quantities of the protein which it was 'programmed' to produce by the DNA transplant. This is well illustrated in Figure 6.1.

The most important breakthrough in the development of rDNA was the discovery by Boyer and Cohen of the use of restriction enzymes, which are in effect chemicals which enable the chain of DNA to be cut, inserted and accepted by its foreign host.[1] This discovery came in 1973 and its significance was immediately recognised. Indeed, the early biotechnology firms were founded expressly to exploit this technique and by the end of the 1970s they had successfully cloned by this method the genes for insulin, interferon, human growth hormone and urokinase—all proteins with therapeutic applications. It has, however, taken them considerably longer than many forecast to move from laboratory cloning to production on a sufficient scale to enable clinical trials to proceed, but all of these early products are now subject to clinical trials and many will be appearing on the market within the next few years. To date, only one, Genentech's insulin, has received US Food and Drug Administration (FDA) approval and is being marketed in the United States and Europe.[2] This managed to pass through the regulatory procedures very speedily, but insulin has been a long-established treatment for diabetes and it is not clear that other genetically engineered products will have so easy a ride.

The other technique of genetic engineering is cell fusion. Where a large part of the genetic message is to be transferred, cell fusion

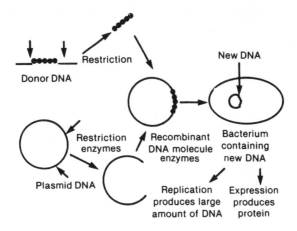

Restriction enzymes recognize certain sites along the DNA and can chemically cut the DNA at those sites. This makes it possible to remove selected genes from donor DNA molecules and insert them into plasmid DNA molecules to form the recombinant DNA. This recombinant DNA can then be cloned in its bacterial host and large amounts of a desired protein can be produced.

SOURCE: Office of Technology Assessment. (1982)

Figure 6.1 Recombinant DNA: the technique of recombining genes from one species with those from another.

may be a preferable technique to rDNA. In effect it enables the fused cells to exchange genetic information and produce hybrid cells which combine the characteristics and/or functions of the original cell. It has potential application both to plant and animal cells but, to date, most experiments on plant cells have been disappointing. Its most dramatic success is in the production of monoclonal antibodies. These are derived by fusing mammalian cells producing a particular antibody with a type of cancer cell, called myeloma. (Monoclonal antibodies are sometimes called hybridomas.) The resulting fused cells are able to produce pure antibodies on a very substantial scale. Until this technique was developed, the main source of antibodies had been extraction from animal tissue which was both expensive and liable to high levels of impurities.

The great advantage of monoclonal antibodies lies in their specificity. Each antibody attaches itself to a specific antigen; therefore 'tagged' antibodies can be used to identify the presence of a particular antigen, to help in separation processes or to deliver drugs to a particular location within the body. To date the main usage of monoclonal antibodies has been in diagnostic processes and purification. Their use as a drug delivery system, which requires injection into the body, obviously has to meet the same toxicology standards as other drugs, and this inevitably slows down their adoption for such purposes.

These developments in genetic engineering are probably the most exciting and significant developments since the New Stone Age experiments which led to the breeding of animals and plants for a settled, non-nomadic form of agriculture. Since then nature has constrained further experiment to the slow and uncertain process of breeding mutant strains. Genetic engineering offers the potential for change by a much surer and quicker route. What should not be forgotten, however, is that genetic engineering by itself can achieve little beyond creating a novel (unnatural) genetic combination in a laboratory dish. It is the *convergence* of developments in microbiology, biochemistry, genetics and chemical engineering that makes the new biotechnology so powerful.

The industrial impact of biotechnology

Table 6.1 summarises the likely industrial impact of biotechnology. A glance down the first column gives some idea of the breadth of its impact—from pharmaceuticals to agriculture; from the chemical industry to electronics. The final column, however, also makes clear how uncertain the date of commercialisation (i.e. general availability on the market) still is for many products. Many of the new therapeutic protein products, such as interferon or urokinase, will become available in the course of the next five years, but other developments, for example the development of new plant varieties via the genetic manipulation of plant genes, may well not come to fruition until the twenty-first century. Other developments, such as biomass energy, represent technologically feasible production methodology which is at present not economically viable in most situations.

The diffusion pattern of biotechnology is in fact typical of that of any new technology. The early markets to be tapped are the high value added markets where high profits can compensate for high risk. Hence the intense activity currently centred on the pharmaceutical and health care markets. In one sense, however, biotechnology is not typical of other new technologies. The previous section emphasised how biotechnology is currently meshing together first, second and third generation techniques. Much that is currently going on in

Table 6.1 The industrial applications of biotechnology: summary table

Industries affected	How affected	Examples of new products	Current world markets (where known)		Projected world markets (range of estimates)		Projected date of commercialisation
Column: 1	2	3	4		5		6
				$m	Year	$m	
Human health care/ pharmaceuticals	New protein therapies	Interferons Human growth hormones Insulin	Interferon HGH Insulin	59b 71a 200a	1990 1990	2,900d–12,600e	1990 (Interferons—Cetus/ Shell Genentech/ Roche)
	New vaccines	Herpes vaccine Hepatitis B	All vaccines	406a	1995	18,000f	1990 (Cetus—vaccines)
	New diagnostic tests	Via monoclonal antibodies	All diagnostics Immunoassay	4,000a 812a	2000	5,000g–43,000h	Coming on market now
	Improved methods of producing antibodies	Penicillin Cephalosporins Tetracyclins	All antibiotics	9,135a			Experiments under way to improve antibiotic yields
	New methods of drug delivery						
Animal health care	New vaccines	Foot-and-mouth Rabies vaccines					1987 (Foot-and-mouth vaccine—Genentech/IMC; Wellcome)
	Animal growth hormones	Bovine growth hormones					1986 (Bovine growth hormone Genentech/ Monsanto)
	Diagnostics and drug delivery						

Industries affected	How affected	Examples of new products	Current world markets (where known)	Projected world markets (range of estimates)	Projected date of commercialisation
Column: 1	2	3	4	5	6
			$m	Year $m	
Agriculture	Tissue culture techniques for plant reproduction	Lettuces Christmas trees Palms, etc.		1990 4,500i	Now in common use
	Plant genetics	Plants which fix own nitrogen Drought resistant strains	Seeds 3,045a	2000 50–100,000g	1984 Calgene/Monsanto genetically engineered strains resistant to Monsanto herbicide. Unknown how long before general commercialisation.
	Biological pesticides and fungicides	Pheromes Fungal sprays	Insecticides 500a		Beginning to come on to market
Food and drink industries	Tissue culture —growing nutritious parts of foodstuffs in fermenter	Extraction of drugs and essential oils			Experiments under way
	New foodstuffs	Single cell protein High fructose syrups Aspartame (sugar substitute)	High fructose syrups 1,300a Aspartame 4.4c		In current production

	New preservatives, additives, etc.	Xanthum gum (filler)			In current production
	New processing techniques	Automated, continuous process bioreactors, etc., use of biological sensors			Coming on-stream now
		Improved strains of yeast, rennet, etc.	Baker's yeast 1,096[a] Rennin 40[b]		Genetically engineered rennet now in use
Fine chemicals	Biological methods of production substituted for chemical synthesis	Production Vitamin C Amino acids Enzymes	Citric acid* 426[a] Vitamin C* 406[a] Amino acids* 1,522[a] Enzymes* 400[a]	1990 Chemicals—5,000[d] Fuel—9,400	Many processes already use biological roiute; others not currently economic, but this may change as a result of (a) relative fall in cost of starch/sugar substrate *vis-à-vis* pharmaceutical food stock, or (b) yield improvement via genetic engineering
Bulk chemicals and fuel	Biomass production of methane/ethanol, and their use as a feedstock for production of bulk chemicals	Acetone Ethylene glycol Glycerol Ethanol Methanol Biodegradable plastics	Methane* 12,572[c] Aliphatics* (other than methane) 2,737[c] Aromatics* 1,251[c] Inorganics* 2,681[c]	2000 Chemicals—9–10,000[d] Fuel—16,400	Biomass production ethanol/methane already common, but unlikely to become more than marginal source of fuel this century and most unlikely to become source of feedstock for bulk chemicals

Industries affected	How affected	Examples of new products	Current world markets (where known)	Projected world markets (range of estimates)	Projected date of commercialisation
Column: 1	2	3	4	5	6
Waste management and pollution control	Biodegradation of domestic and industrial wastes.	Use of new techniques of aerobic and anaerobic digestion.			Techniques already widely practiced, but potential for expanded use substantial.
	Biological methods of purification and detoxification	Genetically engineered microbes for faster degradation Use of monoclonal antibodies			Little or no use of genetic engineering in this sector yet.
Oil extraction	Enhanced oil recovery	Xanthum gum Microbial injection with molasses or other feedstock			Already being tried out. Still in experimental stage.
	Cleansing oil slicks, tanks, etc.	Bacteria injection			
Mineral extraction	Microbial ore leaching	Copper production via microbial methods			10%US copper extracted via microbial methods
Bioelectronics	Use of biological macro molecules as sensors and activators	Use of biosensors in bioreactors The biological chip			Biosensors in increasing use, but more sophisticated developments leading to biological chip not foreseen until next century

biotechnology—for example developments in waste management systems or in mineral ore leaching—are essentially second generation techniques applied and extended to specific areas of interest. General publicity surrounding biotechnology has led to renewed interest in the potential for biological processing, has often shown the considerable unutilised potential for the application of second generation techniques (i.e. the use of natural microbes).

Columns 4 and 5 of Table 6.1 are relatively incomplete and a word of warning is necessary. Many of the new products coming on to the market via biotechnology—for example the interferons—are highly specific and do not themselves find a place in any official statistics. Current market values for these products are derived from consultants' reports or the trade press and are highly speculative. Nor is the list given in column 3 exhaustive; and the values given in column 4 are similarly illustrative, aiming to give the reader a sense of relativities rather than a comprehensive listing of markets. Column 5 is even more limited—indeed it is included to illustrate the sheer range of the numerous predictions currently being made about biotechnology markets in the next decade—a range so wide as to be meaningless. The fact is that at the moment nobody knows, for example, whether interferon will prove to be the wonder drug it was at one time made out to be, or whether it will only have very limited and highly specialised use.

Table 6.1 speaks for itself. However, it is perhaps worthwhile adding a word of explanation about the likely impact of biotechnology in the main areas affected.

Human health care
The important factor here is that biotechnology, although it will bring new products in the form of new protein drugs and new vaccines, is really part of a much wider revolution which is taking place in medicine. Greater understanding of the body's physiological

Notes to Table 6.1.
    This table summarises data given in greater detail in Sharp (1985), Chap. 3. Sources for statistics given in Column 4 are as follows:
    a Dunnill and Rudd 1984.
    b OTA (1981).
    c OECD (1982).
Estimates quoted in Column 5 are all taken from US Government (1982), Table B14a. The specific estimates quoted are as follows:
    d TA Sheets & Co.
    e Business Communications Co.
    f Predicasts.
    g Policy Research Corporation.
    h Robert S. First & Co.
    i Strategic Inc.
  * Current production derived mainly from petrochemical feedstocks.

processes, and in particular the functioning of the immune system, is bringing far greater precision generally to medicine. In pharmacology, it has already brought a shift from general blockbusting therapies to drugs targeted far more precisely to specific problems. Biotechnology, and specifically genetic engineering, has enabled a whole generation of new products to be synthesised which before were either non-existent or virtually unobtainable. The other factor to bear in mind is that, with the exception of Eli-Lilly's insulin, no genetically engineered product has yet been marketed for human use. Many of these new products are currently undergoing clinical trials and as column 6 makes clear, they should be making their appearance within the next five years. But problems of scale-up and marketing are often greater than envisaged. Engineered strains have shown a tendency to revert to their origin, when bred over many generations, and satisfying safety and patent criteria often poses problems. The most flourishing area is that of diagnostics, where, because the tests are all done in a test-tube (or equivalent), the stringent safety rules do not apply. Monoclonal antibodies have opened up a whole new field of diagnostic testing and provide a useful source of cash flow in an area which is still remarkably short of cash products.

Animal health care

Many of the new products being developed for the human health market are also applicable to the animal health care market. As examples, recombinant DNA will ease the problems of producing vaccines for some of the most virulent viruses, for instance foot and mouth disease; growth hormones will mean more output (meat) for input (feedstuffs), although ethical issues may be raised by their use. Meantime, much is already being achieved by the 'old fashioned' methods of cross-breeding and selection, aided by new fertilisation and implantation techniques.

Agriculture

This is likely to be the area of greatest impact for biotechnology in the long run. Developments in animal husbandry have already been mentioned. In cereal and plant propagation, two developments will have major effects. One is tissue culture. This is already being used to generate plant stock for horticulture (for example lettuces) and forestry (Christmas trees and other conifers, oil palms) which saves a great deal of time in crop propagation and ensures a largely uniform and disease-free crop. Tissue culture is likely to be extended to many vegetables (including potatoes), while hybridisation techniques offer the possibility of new plant varieties. Tissue culture may also be used to culture those parts of plants necessary to provide particular nutrients, such as essential oils and flavourings, and this means that factory production effectively replaces agriculture (as has already happened in sugar with the development of high fructose syrups and aspartame).

The other important development is the possibility of genetic engineering in plants. To date rDNA techniques for plants have met with very limited success, plant genes being much more complex than micro-organisms, but the hope is that genetic manipulation of plants will enable genes for various traits (for instance drought resistance or nitrogen fixation) to be transposed into plants. A small American biotechnology company—Calgene—has successfully cloned a gene for resistance to the Monsanto herbicide Round-up and transferred it into cotton plant cells (*New Scientist*, 1983), but it may be many years before the technique is generalised. Interestingly, however, companies such as Shell and Monsanto are buying up seed companies (Sargeant, 1984).

Food and drink industries
The impact here will come from: (a) new ways of producing old products—e.g. tissue culture techniques for producing flavourings, etc; (b) new products—e.g. single cell protein, high fructose syrup or aspartame (another sugar substitute) and (c) improved techniques of production—e.g. quicker-acting yeasts, use of biosensors, etc. to improve processing methods and controls.

Fine chemicals
These are largely used in food and pharmaceutical industries and for specialised uses in other industries (e.g. textiles and paper). Currently many are made via chemical synthesis or extracted from plant or animal tissue at high cost. Tissue culture techniques and/or the biological route to manufacture may reduce production costs and open up new uses, but Europe will find it hard to compete with American and/or Japanese producers as long as high cereal/sugar prices mean high feedstock costs. At present two European firms, Novo Industry (Denmark) and Gist Brocades (Holland) dominate the world enzyme market, while Japan dominates the amino acids. But Japan is making a major effort via its food manufacturing firms to get into these markets, and in general is applying biotechnological techniques far more widely than European firms (Sharp, 1985).

174 Margaret Sharp

Bulk chemicals and energy
Currently, some of the simpler commodity chemicals such as butanol, acetone and acetic acid are manufactured via fermentation, but petroleum provides the main feedstock for bulk chemicals and unless the cost of producing ethanol (hence ethylene, etc.) is very markedly reduced *vis-à-vis* petroleum or methane, this is unlikely to change. At present ethanol production is only economic where there are major sources of biomass to hand and foreign exchange is at a premium (e.g. Brazil). Great strides, however, are being made in producing mobile small-scale fermenters for ethanol production from agricultural and domestic wastes in the developing world. These could well become an important, continuing but marginal source of energy. Meanwhile, the bulk chemical industry is shifting to locate close to cheap (OPEC) sources of methane and the Western European industry is looking more and more towards markets in speciality chemicals and pharmaceuticals.

Waste management and pollution control
The anaerobic digestion of waste materials is already widely applied in the treatment of sewage, animal slurries, etc. Biotechnology offers considerable potential for raising process efficiencies and/or maximising by-product output, but it is a potential that will not be realised unless public authorities (a) invest in the facilities and (b) maintain a tough anti-pollution stance.

Other industrial applications
Other applications worth mentioning are *enhanced oil recovery*, when carefully chosen bacteria plus some cheap source of carbon are injected into oil wells to multiply, increase the pressure and force oil out (though no satisfactory bacteria has yet been found); *mineral ore working*, using bacteria to leach out the mineral content of low-grade ores, which can be extracted by processing the leach solution (currently some 10 per cent of American copper production comes from this source); and *bio-electronics*, a young but growing field in which biochemical reactions are used to replace the circuitry of electronics. Biosensors are increasingly used in computerised control systems in process industries; the biochip has yet to come.

Country-based developments

It is clear from the previous section that biotechnology will affect a diverse and complex series of activities. In this it is, like microelectronics, a generic technology, its impact spilling over into a large number of activities that will gradually, over time, effect major changes in our ways of doing things. Like microelectronics, biotechnology is at root a *process* technology (i.e. to do with ways of *doing* things), rather than a *product* technology, although as a process technology it opens up ways to create new products and, in its early phases of development, much interest focuses on these new products.

In studying country-based activities in biotechnology, it is worth bearing in mind this longer-run scenario and remembering that, as a new generic technology, 'the new biotechnology' is very young and has a long way to go. By analogy with microelectronics, biotechnology is still in the age of the transistor. Much will happen in the next twenty-five years and it is not yet apparent what the course of events will be.

In this section we review country-based activities in biotechnology. Although it is primarily focused on Europe, European experience has to be seen in the context of world-wide developments, where two countries, the United States and Japan, dominate. The section begins therefore with a brief discussion of the development of biotechnology in these two countries.

THE UNITED STATES

The United States is widely acknowledged to be the world leader in two important aspects of biotechnology—recombinant DNA and plant genetics. Its superiority is founded upon a fast-moving intellectual base which has been rapidly able to translate itself into the commercial market-place. In part, as with microelectronics, this intellectual superiority derives from the weight of federal funding. Federal support for the National Institutes of Health, for example, amounted in total in 1982 to some $3.5bn, of which $380m supported work directly related to biotechnology (OTA, 1984, p. 309). Federal and state spending on agricultural research is estimated at $1.5bn, of which $35m went to biotechnology; the National Science Foundation supported biotechnology to the tune of $52m, while the Department of Energy research on biomass added another $36m. This makes a grand total of American government research spending on biotechnology of $510m for 1982, some ten times that of the next highest spender—France (see p. 187 below).

Add to government research spending the private funding for cancer research in the United States, much of which helps fund leading institutes such as the Wistar Institute in Philadelphia and the

Harvard Medical School, and it is not difficult to understand why the United States leads the world in the fields of genetic engineering, immunology, and molecular biology. The facilities available at the universities and institutes have inevitably also proved a magnet to some of the best brains elsewhere in the world: Britain, France, Germany, Japan and the Third World have all lost many of their brightest researchers to American institutions.

Where the United States succeeds so well is in translating this intellectual base into commercial practice. They have created a new breed of intellectual entrepreneur who combines dynamism with drive and flexibility. Part of the explanation is cultural—the 'frontier-spirit' mentality in which everyone is a potential millionaire—and the institutions which have grown up around this culture help to further its cause. There is a venture capital market willing to take on high risks (in return for high rewards), a securities market and tax laws which encourage personal investment through tax write-offs, stock exemptions, etc.

The biotechnology companies that sprang up on the East and West Coasts during the late 1970s were very much creatures of this culture (see Table 6.2). The remarkable breakthroughs of molecular biology and their potential impact on the world of medicine, agriculture and energy caught the public imagination. The academics were prepared to tout their expertise, and the public to support them while the multinationals provided a market for their skills. In the early days (Genentech, for example, was founded in 1975, only two years after the Cohen/Boyer breakthrough on restriction enzymes), it was very unclear whether rDNA would succeed; the problem of

Table 6.2 Emergence of new biotechnology firms, 1977–1983

	Number of new firms founded	Equity investment by established US firms ($m)
1977	3	2
1978	4	32.5
1979	6	22.25
1980	26	77.91
1981	43	78.20
1982	22	119.30
1983	3	(80.00) est.

Source: OTA (1984), pp. 93 and 103.

scale-up and development were untackled and the issue of patent, safety and ethical controls was an unknown. In such circumstances it suited the major companies to bide their time. A $10m research contract with Genentech, Cetus or one of the other companies was a good way to hedge bets.

The biotechnology bubble was in part, however, a self-induced phenomenon. Lucrative research contracts from the major American companies enabled the small specialist firms to go first to the venture capital market and subsequently for share flotation. This in turn encouraged them to 'talk big' about their potential achievements. Major new products, particularly in the health care field, were paraded loudly through the conference circuit and the press conference. The major companies not only placed research contracts with these firms, but also bought equity stakes (see Table 6.2). A new form of investment was developed—the equity partnership—which brought money in to fund specific development but (unlike research contracts) gave both partners shares in the equity on discoveries. The tide turned in 1982. The problem was the lack of products to yield cash income and the (lengthening) estimates of how long it would take to get new products to market. As the problems of scale-up, downstream processing and marketing loomed larger, so the hand of the larger company strengthened. This is well illustrated in Table 6.2 which charts the emergence of the new biotechnology firms in the United States with the peaking of new establishments in 1981 and of investment by established firms in 1982.

Nevertheless, it is important not to write off too quickly the new biotechnology companies of the United States. Although there has been some thinning out amongst ranks of new establishments, there has not been the mass of failures that some have predicted. There seem to be a number of reasons for this. First, they continue to perform an important intermediary role between academic science and the commerical world—indeed, these companies still provide an amazing source of creativity for biotechnology. Many scientists prefer working for a small, flexible company which is often almost an extension of their academic laboratory, rather than for large bureaucratic organisations. Therefore, although the new biotechnology firms no longer find themselves in the 'easy money' era of 1980, there remains a continuing flow of research contracts from the large firms. Research contracts, however, are less satisfactory than equity investments: it is the large company that will make profits from successful research. Hence the larger of the new companies (Genentech, Biogen, Cetus) are putting much emphasis on bringing their own new products to market. Moreover, while American firms may be loosening their links, foreign companies are strengthening their ties with these companies. A second wave of investment into these companies has come from Japan in particular, and Europe. Finally, quite a number of the companies have moved from research into supplies for, as biotechnology has flourished, so has its 'service' sector which provides specialist equipment, chemicals, etc. Many small firms have also moved into diagnostics where developments using monoclonal antibodies have provided products with cash flow which they so badly need.

It is worth highlighting one further feature of biotechnology in the United States: its bias towards pharmaceuticals and agriculture. Table 6.3 shows the proportion of American biotechnology firms in 1982/3 engaged in different areas of research.

Table 6.3 Proportion of firms in the United States pursuing applications of biotechnology in specific industrial sectors, 1982/3*

	%
Pharmaceuticals	62
Animal agriculture	28
Plant agriculture	24
Speciality chemicals and food	15
Commodity chemicals and energy	15
Environment	11
Electronics	0.03

* Totals to more than 100 per cent because many firms in sample were involved in more than one sector.
Source: OTA Report (1984), p. 71 and Appendix D.

This bias is not surprising. In the first place, the pharmaceutical industry is a high risk, but high return, sector: the firm that succeeds in being the first to introduce new drugs makes very high profits in the first few years, until imitators can catch up. Biotechnology, or rather genetic engineering, opened up a new relatively cheap route into pharmaceuticals—or so it seemed. And the new entrants swarmed into the lucrative market, only to discover that the research phase was but the beginning of launching a new drug. In the second place, it reflects the bias of federal funding which goes predominantly to health and agriculture. With the exception of the now defunct gasohol programme, little public funding has gone into the chemical or chemical engineering side of biotechnology.

Table 6.4 R&D budgets for biotechnology from some leading American companies, 1982

	$m
Schering Plough (pharm.)	60
Eli-Lilly (pharm.)	60
Monsanto (chemicals)	62
Du Pont (chemicals)	120
Genentech (new biotech. co.)	32
Cetus (new biotech. co.)	26
Genex (new biotech. co.)	8.3
Biogen (new biotech. co.)	8.7
Hybritech (new biotech. co.)	6.0

Source: OTA Report (1984), p. 74.

Table 6.4 illustrates the extent to which the major companies in the United States are now investing in biotechnology research and development. Although obviously illustrative rather than comprehensive, when firms as powerful as Du Pont and Monsanto put resources of $60–$100m behind an area of research, it indicates the importance they attach to its development. The table also illustrates how even the largest of the new biotechnology companies —Genentech—invests a great deal less than the big multinationals.

JAPAN

Unlike the United States, Japan's strength in biotechnology lies in the fermentation industry and it is the largest corporations which are developing and exploiting these strengths, rather than small companies. Japan admits that it is weak in rDNA and cell fusion techniques. When these techniques were being developed in the United States in the 1970s, Japan was slow to recognise their importance and further impeded their application by imposing extremely strict safety rules on any experiments. The result was that, by the end of the 1970s, Japan lagged some five to six years behind the United States. Since then, it has been making a determined effort to catch up, through link-ups between Japanese and American companies, particularly with the larger start-up companies, and a substantial programme of training doctoral and post-doctoral students abroad, mainly in the United States.[3]

The primary executive agency for 'big science' in Japan is the Science and Technology Agency (STA). As far back as 1973, the

180 Margaret Sharp

STA established a specialist committee for biotechnology (Committee for the Promotion of the Life Sciences) and in 1981 its budget for research amounted to $210m, although only $24m of this went to strictly defined biotechnology (Rogers, 1982). The STA's research programme concentrates on medical aspects of biotechnology (particularly genetic engineering, and it is building a new facility for this in Tsukuba Science City) and longer-term research projects on protein synthesis and fully automated bioreactors.

The Ministry of International Trade and Industry (MITI) represents the industrial interest in biotechnology, and as might be expected, its emphasis is to a greater extent on the practical application of new techniques. It is responsible for the Fermentation Research Institute at Tsukuba Science City. In 1981 it declared biotechnology to be 'an industry of the future' and inaugurated a $110m ten-year programme of research and developments, with three main areas of interest: bioreactor development ($43m); rDNA (particularly its application to industrial processes—a further $43m budget), and large-scale cell culture ($24m). Some forty Japanese companies are joining MITI in this programme through the linked research association—companies such as Mitsubishi Chemicals, Mitsui, Kyowa Hakko, Ajinomoto, Takeda, Sumitomo. MITI also has an interest in biomass energy with a budget of some $7m per annum (Rogers, 1982).

Considerable rivalry exists in Japan between the various ministries. Besides MITI, both the Ministry of Health and the Ministry of Agriculture are funding research programmes in biotechnology, each of approximately $10m a year (OTA, 1984, p. 317). The Ministry of Health is responsible for the pharmaceutical industry and is using the biotechnology programme as an attempt to reinvigorate the Japanese pharmaceutical industry which has been characterised by small firms with little research orientation (*The Economist*, 1981 and 1984). Putting the budgets of the various ministries and their agencies together gives a figure of approximately $60m in 1982/3 spent by public agencies in Japan on promoting biotechnology (Sharp, 1985, Chap. 4).

Japan's real strength, however, lies in its company sector, both in the number of companies showing awareness of and involvement in biotechnological activities, and in the resources devoted to R&D in this area. A 1982 survey by MITI of 200 corporations showed 157 with an R&D programme in biotechnology already under way (JETRO, 1982). Total research expenditures of the firms involved amounted to Y747.8bn ($203m), with 27 per cent of this being concentrated in the food industries, 20 per cent in the drug industries and 20 per cent in the chemical industries. In the drug industry (not perhaps surprisingly) interest focused on rDNA, cell fusion techniques and monoclonal antibodies, while in the chemical sector, enzyme technology and downstream processing gained concentrated effort. The food industry showed interest in fermentation technologies and downstream processing, although a surprisingly large number of firms claimed activity in the rDNA field.

Aware of their deficiencies in rDNA and cell fusion, the Japanese drug industry has been actively seeking links with foreign companies and in particular have sought links with the major American drug companies and some of the new specialist biotechnology firms. But it is the application of genetic engineering to its traditional strengths in food and fine chemicals that is exciting the Japanese. Ajinomoto researchers are set to improve amino acid production techniques using genetic engineering, claiming to have cut costs of production by as much as 90 per cent. Mitsubishi Chemicals already reputedly devotes one-third of its research budget to biotechnological research and plans to convert 20 per cent of its production to bio-processes by 1991.

It is difficult to know quite how formidable a challenge this adds up to. Traditional Japanese strengths in fermentation, combined with industrial foresight, have already given them economic dominance of the world amino acid market and considerable strengths in other fine chemical and enzyme areas. Current research programmes are reinforcing these strengths and this puts Japan in a very strong position to mount a major challenge in the fine and speciality chemical markets—markets to which the major European chemical companies are currently looking as Middle East production gradually takes over their bulk chemical markets. Whether Japan will also be able to break into world pharmaceutical markets is more open to question. Their drug industry remains weak and fragmented and it is by no means clear that any of their firms are in a position to surmount the major barriers that protect the international drug industry. Nevertheless, biotechnology is shaking these barriers, and the Japanese are determined to be ready to seize the opportunity.

EUROPE

The pattern of biotechnology development in Europe is somewhat closer to the Japanese than to the American experience. Although, as we shall see, substantial differences exist between European countries, in general it has been governments and the large corporations who are providing the lead, the small 'start-up' company playing a very small part. Like Japan, Europe also has traditional fermentation interests and, in Germany at least, this led to early involvement in enzyme immobilisation techniques, and considerable interest in bioreactor development. Unlike Japan, Europe has a very strong pharmaceutical industry, built initially upon the chemical expertise of Germany and Switzerland, but the post-war advent of penicillins shifted the focus to drug development and opened the door to American multinationals.

Looking at Europe as a whole, therefore, the intellectual base for biotechnology is strong and there are plenty of firms capable of developing the technology. The main issue is that of bridging the gap between academic research and industry in the absence of either the American-style academic entrepreneur, or the close relationship of the Japanese conglomerates with the state and research institutes.

Tables 6.5 to 6.7 give some indication of the activity in biotechnology in the different European countries *vis-à-vis* the United States

Table 6.5 American biotechnology* patent grants, January 1963–December 1982

	Cumulative total 1963–82		Grants in 1982	
	No.	%	No.	%
Total patents granted of which	3,381	(100)	372	(100)
United States	1,712	(51)	211	(56)
Other	1,669	(49)	161	(44)
by country				
Japan	805	(23)	70	(18)
France	105	(3)	12	(3)
W. Germany	203	(6)	17	(5)
United Kingdom	188	(5)	25	(7)
Switzerland	46	(1.5)	2	(0.5)
Netherlands	35	(1)	2	(0.5)
Sweden	41	(1)	4	(1)
Denmark	38	(1)	5	(1.5)
Italy	44	(1.5)	4	(1)
Belgium	21	(0.5)	—	

* Patent categories included are: enzymes *per se*, immobilised enzymes, genetic engineering and mutation, tissue culture, starch hydrolysis and amino acids.

Source: Derived from data in OTA Report 1984 and Patent Profiles Update 1982.

and Japan. Table 6.5 details patent activity and gives a clear indication of the dominant position of the United States and Japan in this field of activity. Patent statistics in biotechnology need, however, to be interpreted with care, particularly in the current state of uncertainty about the patentability of biotechnology products and processes. Japanese firms tend to patent extensively, while many European companies go to the opposite extreme and prefer to rely upon in-house loyalties rather than reveal strategically important areas of research. Patent data is not therefore an entirely reliable guide to activity. For what they are worth, the figures suggest that, among European countries, West Germany and the United Kingdom are the most active participants, followed by France, with countries such as Switzerland, the Netherlands, Italy, Sweden and Denmark all making some contribution.

Table 6.6 Government spending in biotechnology programmes—European governments compared to the United States and Japan

	1982/3	$m
	Own currency	
United States (FY 1982/3)	$510.5m	510.5
Japan (1983)	Y14,761m	59.3
W. Germany (BMFT 1983)	DM 96m	37.7
France (1983)	890m frs	117.0
United Kingdom (FY 1982/3) including MRC underpinning research	£28.9m (£45.9m)	43.9 (69.8)
Netherlands (1982)	NG1 25m	9.4
Italy (1983)	L. 7bn	4.6

Source: Sharp, 1985, Table 9.2.

Tables 6.6 and 6.7 fill out the background on public spending on biotechnology and related activities. The detailed figures for specific spending on biotechnology are derived from country-based reports and do not always compare like with like. More reliable are the R&D expenditures statistics quoted in Table 6.7 and derived from EEC data. Together, the two tables show France as the country currently devoting most resources to promoting biotechnology, but West Germany as the country committing most resources to the broad research base in medicine and agriculture upon which biotechnology

Table 6.7 Publicly funded research spending in areas broadly related to biotechnology, 1982

	Total	Medical/ Biological Science	Agriculture	Food and drink	Environment
					ECU million
W. Germany	1503.7	988.6	331.2	42.3	140.5
France*	842.8	535.5	206.7	42.4	58.2
Britain	598.6	233.8	330.5	21.5	65.6
Italy	425.5	234.7	113.3	11.9	65.6
Netherlands†	321.4	180.4	95.1	n.a.	45.9
					$ million
USA (1982–3)	5202	4119	846	n.a.	237

1 ECU = $0.97 in 1982.
*1980 figures derived from Cantley (1983).
† Incomplete total.
Sources: Sharp, 1985, Table 9.3.

draws so heavily. It is worth remembering that salaries in France and West Germany are between one and a half to two times salary levels in the United Kingdom and Italy and the seemingly low spending of these latter two countries is not as poor as appears. It is interesting to note, however, that the United Kingdom is the only country to spend more on agricultural research than on medical and biological research.

The details on public funding of research are discussed separately below in the country sections.

WEST GERMANY

West Germany's traditional intellectual strengths in chemistry and the country-wide interest in beer and wine make it a 'natural' for biotechnology. Indeed, one of the outstanding features of German experience in this area was the early recognition of its potential importance, matched by government research funding from the early 1970s onwards. Given these strengths and the strong government backing, achievements have been somewhat disappointing.

The intellectual base for biotechnology in Germany is extremely strong. There are four main 'centres of excellence' at Cologne, Heidelberg, Munich and Berlin, where university facilities are reinforced by a Max Planck Institute (MPI) and where there are also strong institutional links with one of the major pharmaceutical/chemical firms.

There is also widespread interest and involvement in many universities, for example, Tübingen, Hanover, Bonn, Aachen and Frankfurt.

The main centre for biotechnological research is, however, the Gesellschaft für Biotechnologische Forschung (GBF) at Braunschweig. Founded in 1968 by the Volkswagen Foundation, it was taken over in 1975 by the Federal Government and is now funded 90 per cent by a Federal Government Agency, the Bundesministerium für Forschung und Technologie (BMFT), and 10 per cent by the Government of Lower Saxony. Its research programme includes genetic engineering and microbial techniques, but its strengths lie in process technology, including the development of bioreactors and the problems of downstream processing and control. It also has a major interest in enzyme technology, and in cell culture mechanisms. Its research budget in 1982 was DM31.6m ($13m), to which the BMFT contributed DM28m ($11.5m) (Sharp, 1985, Chap. 5).

In the Federal Republic a key role in the development of biotechnology has been played by DECHEMA, the chemical plant trade association. It provided the initial stimulus to interest in biotechnology as long ago as 1968, wrote the first major report on the subject for the BMFT (DECHEMA, 1974), and has since remained the focal point for the German 'biotechnology lobby'. Overlapping membership of its own committees and that of the advisory committee for the BMFT and the DFG (German Science Foundation, which advises on the allocation of funds to university science research) means a considerable convergence of view between these three institutions.

Government and trade association interest helped to stimulate early industrial involvement in biotechnology. Hoechst moved into single cell protein; Boehringer Mannheim developed interest in enzymes and enzyme technology; Bayer moved into biological pesticides and herbicides; Dynamit Nobel worked on amino acid development from enzyme materials; Degussa worked on scaling-up enzyme reactor systems. Interest in bioreactors for one purpose or another was widespread: Boehringer Mannheim was one of the main recipients of BMFT grants for bioreactors; Bayer developed a new deep-shaft sewage system, and Messerschmidt was interested in using methane from waste matter for fertiliser development. Many of these projects were 'seeded' by BMFT money, which also encouraged collaborative research with universities and specialist institutes.

The interesting feature of these German developments is the extent to which they mirror Japanese experience, with attention focused on enzymes, enzyme technologies and bioreactors. Whereas in Japan it was the food industry, and particularly firms like

Ajinomoto, who took the lead in pushing into large-scale production of amino acids via fermentation, in Germany the food industry, a fragmented industry, has played a negligible role. Although Degussa, the Frankfurt chemical firm, has considerable interests in amino acids, no European firm rivals the major Japanese producers.

While the role of DECHEMA stimulated the early developments of biotechnology in Germany, it can be argued that its interests as a chemical plant association in enzyme technology and bioreactors have, together with environmental pressures, led biotechnology in Germany up something of a blind alley. Like Japan, while concentrating in the 1970s on developments in enzyme technology, West Germany was bypassed by developments in genetic engineering. A dramatic illustration of this came in 1981 when Hoechst, the world's largest pharmaceutical company, chose to invest DM113m ($50m) over a five-year period in the Massachusetts General Hospital (which has research links with Harvard and MIT), in order to gain access to American genetic engineering expertise, admitting that no institute in Germany combined the level of expertise and the 'lateral thinking' capacity of the Americans.[5] Hoechst and Bayer, of course, both have strong subsidiaries in the United States and through them have been able to keep in touch with developments in that country. Recent estimates suggest that each of these firms is currently spending $60–$70m on biotechnology research (Junne, 1984), which puts them in the same league as Monsanto and Du Pont.

The Hoechst deal with MGH in 1981 both offended and woke up the German biotechnology establishment. The BMFT increased its budget funding for biotechnology from DM40m in 1980 (approximately $22m) to DM63m in 1983 (approximately $25m), putting cell fusion and genetic engineering high on its list of priorities. (An additional DM33m ($13m) goes from the BMFT to fund institutions such as Braunschweig.) Other initiatives were also taken to boost biotechnology. The regional government of Baden-Württemberg put DM30m (approximately $12m) into the construction of a new institute of molecular genetics at Heidelberg which will also be supported by the BMFT (DM13m ($5m) over three years) and the chemical firm BASF (DM4m ($1.7m) over five years). Bayer have subsequently announced support for the Cologne Max Planck Institute and Schering AG have joined forces with the city of West Berlin to put DM40m ($16m) into a new Biotechnology Institute at the technical university (Davies, 1983). Public spending cited above does not include money going into biotechnology via medical research or agricultural research. The Federal Government spent approximately $950m on medical research funding in 1982 and $330m on agricultural research (Eurostat, 1984). A rough rule of thumb is that 10 per cent of these expenditures is on broadly defined biotechnology.

As in Japan, the main thrust into biotechnology has come from the large firms and the German scene is remarkable for the relative absence of the small biotechnology firm. This is partly explained by the traditional relationship between the large chemical firms and the universities and research institutes in Germany: as a scientist from one large chemical firm explained in an interview, 'If we have a research problem we cannot solve, we turn to the universities for an answer'. But it also reflects the areas and timing of the German research thrust. Until 1980/81, the main thrust of research had been in the enzyme/bioreactor area, and was essentially process-orientated. By the time the large German companies moved into genetic engineering, the major uncertainties of the early years had been overcome. Genetic engineering had been shown to be a technique which worked, and the German companies were merely following their major American rivals in committing substantial internal research resources to biotechnology. Moreover, now that the emphasis is moving from the laboratory to pilot plant production, German excellence in chemical engineering may come into its own. In so far as Germany has any continuing weakness, it is probably in the academic field where commentators have remarked upon the surprising failure of universities to meet the research needs of industry and suggested that there is continuing overload of administrative and teaching duties to the detriment of research (Dunnett, 1983).

FRANCE

France woke up to developments in biotechnology relatively late, but made up for this by publishing three reports within the span of two years (Gros, Jacob and Royer, 1979; de Rosnay, 1979; Pelissolo, 1980) and by initiating a bold programme of development aimed at making France the leading European country in biotechnology by the end of the century.

France has a strong tradition of academic research in microbiology and allied disciplines. Two of its research institutes are of world renown—the Institut Pasteur in medical research, and INRA, the Strasbourg-based agronomics institute whose interests include soil microbiology, plant genetics and nitrogen fixation. There are also a number of other research institutes—for example INSERM (institute for health and medical research) and the IFP (energy research) with interests in biotechnology. The Government's plans involve linking these research institutes with the universities to form a major research base. University research in these areas in France has generally been weaker than the research institutes and only three have faculties of world standing—Compiègne (enzyme technology and biochemistry), Toulouse (microbiology, biochemistry and chemical engineering) and Strasbourg (genetic engineering). On the whole, French traditions in applied research are weak, and biotechnology suffered from the compartmentalisation that still persists in universities.

The Government's plans are to mobilise both research and commercialisation around a series of 'pôles de développement'. For example, the Institut Pasteur, along with the other research institutes (CNRS, INSERM and INRA), provides the core team on genetic engineering for pharmaceuticals; the IFP, CNRS and Toulouse University are to concentrate on fermentation technology for chemicals. Similarly, commercialisation is to be led by a group of core firms and their subsidiaries.

One of the biggest problems France faces in pursuing these initiatives is the relative weakness of its pharmaceutical and chemical firms. In pharmaceuticals, no French firm has a tradition in antibiotics. The strongest firm is probably Roussel Uclaf, which manufactures both vitamin B12 and the amino acid methionine, but the firm is now 60 per cent owned by Hoechst which gives it certain advantages (for example, access to Hoechst research and American links), but also sets it somewhat apart from government-sponsored efforts to promote biotechnology. This leaves the major firms in the chemical sector—Rhône-Poulenc and Produits Chimiques Ugine Kulhmann—both suffering badly in the recession and in the midst of a restructuring programme aimed at switching their emphasis towards the speciality and fine chemical segments of the market. The key firm in this area is in fact Elf Aquitaine, the government-owned oil company. With its own interests in biomass energy, and a major subsidiary, Sanofi, in pharmaceuticals, it is spearheading government efforts in the company sector. It has itself put 100m frs ($15m) into a new research centre at Toulouse University; has established two subsidiaries—Elf Bio-Industries and Elf Bioresearch—to develop research and interests in the agri-food sector, and is a minority shareholder in Transgène—the Strasburg-based new biotechnology company, financed initially by Paribas, and which has been a considerable success story.

With food and wine being such traditional industries, one might have expected to find France stronger in the agri-food sector. But tradition also makes for conservatism and the large French firms in this sector have in the past spent little on research. The Government is pushing some of the foremost of these—BSN-Gervais-Danone, Moët-Hennessy and Bel—into taking an interest in biotechnology. The first two are minority shareholders in Transgène.

France's greatest strength in fact lies in its medium-sized specialist firms. Orsan (a subsidiary of the conglomerate Lafarge Coppée) and Eurolysine (taken over in the late 1970s by Gist Brocades of the Netherlands) produce the amino acids glutamate and lysine, and Roquette is one of the world's leading producers of starch-based chemicals and the world's largest producer of sorbitol, the sugar substitute. Lasaffre is a specialist producer of yeasts.

The Mobilisation Plan for biotechnology of July 1982, drawn up by the new Socialist Government, drew heavily on the ideas of the de Rosnay (1979) and Pelissolo (1980) reports. These had argued for a strategy involving Government leadership, with university and industrial participation, and building on France's traditional strengths in agriculture and food production, but seeking to remedy weaknesses in such areas as monoclonal antibodies. Under the Mobilisation Plan, France's share of world trade in biology-based products was set to rise from 7 per cent to 10 per cent, with some forty projects being discussed with the major chemical and pharmaceutical firms where the Government would help with financing. Altogether, the Government was aiming to put something like 400m frs ($60m) per annum into biotechnology, hoping that industry would match this with expenditures of 600m frs ($91m) (Yanchinski, 1982). Recent estimates, which include the cost of public laboratories and general R&D support, put total public expenditure on biotechnology in France at 650m frs ($98m) for 1982 and 890m frs ($117m) for 1983 (Eurostat, 1984).

With the substantial cutbacks in public expenditure since 1983, the more ambitious plans have been trimmed. Moreover, French industry, hit by the recession, has not been able to meet anything like the levels of expenditure projected. Nevertheless, while funding and timetables may slip, the essence of the strategy remains. What the French are trying to do is to force a marriage between a somewhat reluctant industry and a good, but not entrepreneurial, research base. Whether they will succeed is a moot point. In their favour is the French ability to mobilise resources—the centralisation of the education and university system; the tradition of administrative discretion wielded by French civil servants towards industry; and the now extensive nationalisation of the large firms. Against them is the fact that the areas chosen for particular attention—agriculture and food manufacturing—are innately conservative and have traditionally spent little on research. Nor does France have the West German competence in chemical engineering. Moreover, while plant cloning techniques are now well established, for many of the areas where France is seeking to forge ahead, nitrogen fixation, plant genetics, biomass using sugar beet waste and cellulose degradation, the strategy is high-risk and the pay-off long-term. But such a strategy has worked for the French in areas such as nuclear power (Surrey and Walker, 1981) and offshore (see Chapter 7 in this book). It could well take France to a position of considerable strength in these areas by the end of the century.

THE UNITED KINGDOM

Of all European countries, the intellectual climate in the United Kingdom most closely approaches that of the United States. British scientists have played a seminal role in the development of modern biotechnology, particularly in genetic engineering. Cambridge, of course, stands out—from Crick and Watson's work on DNA in the early 1950s to Millstein's work on monoclonal antibodies in the 1970s. But Cambridge does not stand alone: the roll of honour includes for example, Wilkins and Franklin at King's College, London; Alexander Fleming at St Mary's Hospital, London; the Oxford cephalosporin team; Isaacs discovering interferon at Mill Hill; the pioneering of foot and mouth vaccines in the Agricultural Research Council's laboratories at Pirbirght. As in the United States, there are a number of centres where there is a gathering of universities and research institutes to form a relatively large 'biotechnology community' which encourages the cross-fertilisation of ideas so essential to a mixed discipline subject like biotechnology.

Academic research has tended to concentrate at the 'high tech' end of genetic engineering with, at least until recently, less attention being paid to enzyme and fermentation technologies and chemical engineering. This is reflected in the bias of public R&D funding in the areas broadly related to biotechnology—of £342m ($598.5m) expenditures in this area in 1982, 38 per cent went on medical research and 55 per cent to agricultural research (Eurostat, 1984). Recent funding decisions have tried to reverse the balance and new research money for biotechnology has gone disproportionately towards the downstream technologies. But, in a climate of general cutbacks in university finance, substitution is frequent and it is not always easy for new initiatives to achieve their targets.

Contrary to the generally held view, one of Britain's strengths in biotechnology lies in the breadth and depth of its industrial experience and interest in this area, an interest which stretches back over two decades or even longer. In contrast to the United States, but in common with Japan and the continental European countries, this interest comes not from small specialist firms, but from large, well-established firms, many of them multinationals, with substantial R&D departments. Unlike the new American biotechnology companies, they have no particular need to broadcast their activities; on the contrary, given the oligopolistic environment in which most operate, they are often anxious to hide the specific direction of their R&D.

In the pharmaceutical field, the leading British firms are Glaxo, ICI and Wellcome, all of which have traditional interest in biologically-based medicines and strong links with academic research. Seven of the top ten foreign multinationals have major research facilities located in the United Kingdom, and it is the quality of the research base, including molecular biology, that attracts these facilities (Brech and Sharp, 1984, Chap. 4).

Outside pharmaceuticals, one of Britain's greatest strengths derives from single cell protein (SCP) experiments where Shell, BP and ICI pioneered developments based on hydrocarbon sources. Of the three, only ICI have persevered and are now manufacturing their methane-based product, Pruteen, as an animal feed additive using a continuous fermentation process and producing 50,000 tons per annum. Although the economics of the enterprise have been dissappointing, it has given ICI experience in large-scale continous fermentation (Howells, 1982). Tate and Lyle, the sugar group, also have interests in biotechnology, ranging from high fructose syrups and other sweeteners, to carbohydrate technologies (where they have a research link with Hoechst), to alcohol fermentation, SCP, waste product purification and biological pesticides.

In the agri-food field, a number of major firms have interests. Both Shell and Unilever have research interests in plant genetics, and Unilever has pioneered techniques of cloning palm trees, enabling them to concentrate production on hardy and high-yielding varieties. Unilever also have research interests in monoclonal antibodies, both for medical diagnostics and purification (of oils). Rank Hovis MacDougall have developed a SCP foodstuff based on a starch substrate; Dalgety-Spillers, the other major flour millers, have joined forces with Whitbread and Distillers to support yeast research at the Leicester University Biocentre. Cadbury Schweppes have new research laboratories at Reading University; Grand Metropolitan is one of the main shareholders in Biogen and has established a small biotechnology subsidiary which is to be located at the University of Surrey.

Britain's major weakness is in the fine chemicals and enzymes field, where there is no firm to challenge the Danish company Novo or the French Roquette. A number of small companies such as Biozyme and NBL enzymes are now trying to establish themselves in this area, but they remain small and peripheral to the main market. This weakness is reflected in the general lack of interest in the chemical engineering aspects of biotechnology, and for that matter in the process plant industry.

There is, however, an increasingly strong small-firm sector in the United Kingdom. This has developed partly as a result of deliberate attempts, via changes in the tax laws, to create a venture capital market in London. There are now about thirty such companies in the United Kingdom, many not involved in genetic engineering but as specialist suppliers providing the speciality chemicals, enzymes, equipment, etc. needed for biotechnology. Others are closely linked to university departments and aim essentially at selling expertise and consultancy rather than products. The only company which really parallels the larger American biotechnology companies is in fact the government-sponsored Celltech.

Government policy towards biotechnology has been variable. Until 1980 research support had been left to the appropriate research councils (Medical, Agricultural, Science and Environmental), with the bulk of funding being channelled through the Medical Research Council (MRC), although the Science Research Council (SRC—now SERC) had promoted a programme on enzyme technology in the mid-1960s. It was not until the Spinks Report in 1980 (ACARD, 1980) that the British Government began to realise the potential importance of this area, but even then its official White Paper response took twelve months to appear and argued that if biotechnology had so much potential, the market would provide (UK Government, 1981).

The Spinks Committee had argued that in disciplines related to biotechnology Britain had for two decades been a world leader in research and yet was now (1980) in danger both of relinquishing this lead and of failing to exploit the ideas pioneered by its own scientists. Priority therefore lay, first with the reinforcement of the research and teaching base, and second, with making sure that the ideas from research were picked up and developed by industry. In fact, over time, some extra funding has been forthcoming—in the universities for teaching and research facilities and encouragement of collaborative research (by 1984 this amounted to some £5m extra funding per annum); and in industry a £16m ($28m) package, announced in 1982, to be spread over three years and aimed primarily at increasing industrial awareness and helping to bridge the pre-production development gap. Total government funding for biotechnology (including expenditures on underpinning research by the MRC) amounted to approximately £46m ($69.8m) in 1982/3 (Sharp, 1985, Table 7.9).

Perhaps the most surprising outcome of the Spinks Report was the establishment of Celltech, the public sector biotechnology company. Celltech is a hybrid organisation, with a starting capital of £14m, part government-owned and part private sector. Its main function is technology transfer from university and research institute to industry and to this end it has limited first option rights to genetic engineering developments in the MRC laboratories around the United Kingdom.[6] But it does more than merely act as an intermediary, and during its three years of existence has built up a programme based on contract research, licensing and product development, and is a world leader in mass producing monoclonal antibodies. Celltech's parent is the British Government's venture capital agency, the British Technology Group, founded originally by the 1974–79 Labour Government as a government-owned and financed investment group. Besides Celltech, the BTG has set up the Agricultural Genetics Corporation (AGC) on similar lines, to exploit agricultural research in the biotechnology area, and it has a portfolio of some forty other investments in biotechnology, most of them quite small (Celltech, its largest, is only £7m). Government policy is now to sell its stake in such holdings as soon as possible.

In 1980, Spinks had called for a strong, central co-ordination of effort on the part of the British Government. This has not been achieved. There is no grand strategy, identifying growth sectors and putting money and resources into designated projects. In many senses, British Government strategy is very low-key, aimed at stimulating awareness, links between university research and industry and funding pilot projects. Underlying it, however, is a greater degree of co-ordination and thrust than meets the eye. Together, the Research Council initiatives, the DTI package of selective assistance and the BTG investments, plus the stimulus given to the development of the private venture capital markets, adds up to a programme which begins to have both consistency and coherence.

THE NETHERLANDS

The Netherlands has obvious interests in biotechnology. It remains an agricultural country with a major food processing industry and it is the home (at least in part) of Shell, Unilever, and Gist-Brocades, one of Europe's major enzyme producers (and a major producer of penicillin via fermentation). In the post-war years it has developed a strong chemical industry based upon oil and natural gas and it is probably the world's leading country in effluent treatment.

Two reports in the early 1980s (Apeldoorn, 1981; Schilperoort, 1982) emphasised these strengths and pointed to the excellent intellectual traditions in microbiology, biochemistry and process engineering. But as in all European countries, its universities are criticised for being too compartmentalised, for lacking interest in applied research and for being too remote from industry.

In 1982, government funding for biotechnology amounted to 5m guilders (approximately $1.8m), with a further 19m guilders (approximately $7.12m) on university research. Since then, the budget has both been increased (in line with Schilperoort recommendations) and subsequently cut back again. Current expenditure over and above university research spending amounts to approximately 6m guilders a year ($2.3m).

Both the Schilperoort and the Apeldoorn reports recommended a concentration of resources upon a limited number of 'centres of excellence': Leiden for plant genetics; Delft for fermentation technology, and the Agricultural University at Wageningen for animal and food sciences. In the event, some thirty new posts have been created but there seems little effort to concentrate resources—on the contrary, it seems that every university has claimed its due 'share'.

Regulations relating to experiments using rDNA techniques have been particularly strict in Holland and this is now generally regarded as having hindered the progress of genetic engineering in that country. Perhaps for this reason, large-scale use of rDNA technology has hardly begun in the Netherlands, but a large number of companies, including Gist-Brocades, Unilever and Akzo-Organon may well wish to develop their interests on these lines in the future. There are proposals under consideration which would shift the regulations to a more pragmatic basis under which the potential hazards of each project would be considered individually. Current laws, for example, forbid the introduction of recombinant species into the environment.

ITALY

A recent report published by the Federation of Scientific and Technical Associations in Italy (Milanesi, 1984) suggested that, although basic research in biology in Italy compared well with other European countries, the development of biotechnology in Italian firms was relatively poor.

The combination of recent reforms in the organisation of universities in Italy, which have encouraged the training of a new generation of research-orientated scholars, and continued support since the 1950s for biological research from the Consiglio Nationale delle Richerche, mean that basic biotechnological research in Italy is comparatively good.

By contrast, industrial application is disappointing. Three public companies—Farmitalia-Carlo Erba of the Montedison Group, Assoreni and Sclavo of the ENI group—all have some interest in biotechnology, stimulated in part by the Italian Government's desire that Italian companies should develop involvement in this area. But only five private sector companies of any substance have so far shown sufficient interest to commit investment/R&D expenditures—Sona Biomedica (of the Fiat group), Lepetit, Recordati, Soveno Institute and Societa Produzione Antibiotici.

SCANDINAVIA

The experience of the Scandinavian countries in biotechnology is interesting for two reasons. First, although governments are providing substantial and continuing support for biotechnology through their support for basic science at the universities, the main initiatives are coming not from government but from the firms themselves. Second, these firm-based initiatives are creating a strong grouping of companies which might be said to operate almost as a 'regional' grouping, comprehending on the one hand Danish companies (from within the EEC) and on the other Swedish and Norwegian companies (from outside the EEC).

In Denmark, the dominating position is held by Novo Industry—the enzyme and pharmaceutical group—whose profits increased eightfold in the eight years 1973–81. Novo's main product is insulin (originally developed as a by-product from Denmark's agricultural industries) and it has pioneered the chemical modification of the (normally used) porcine insulation to produce an insulin which is extremely close to human insulin. An aggressive marketing strategy in both Europe and the United States has already brought substantial gains at the expense of Eli-Lilly, the world market leader, and its new 'human' insulin is a direct challenge to Lilly's genetically engineered product. Novo is also a major producer of specialist enzymes and enzyme products, a market that has been expanding fast with the growth of biotechnology. It is now the foremost European producer in this market (a position it has captured from Gist-Brocades, the Dutch firm), and expanding fast into new markets, offering a comprehensive catalogue of its own and bought-in products and a strong, efficient marketing service.

Beside Novo, the other leading Danish firm with biotechnology interests is Carlsberg, the lager company, whose brewing interests lead naturally to yeasts and fermentation technology. The Carlsberg laboratories in Copenhagen are a leading centre for studies in cereal genetics.

In Sweden, the leading role in biotechnology is occupied by two pharmaceutical firms, Pharmacia, a subsidiary of the larger pharmaceutical/medical products group A. B. Fortia, and KabiGen, a subsidiary of the state-owned pharmaceutical group, KabiVitrum. Pharmacia is closely connected with Uppsala University and is jointly funding with the Swedish Government an Institute of Cell Research. It is a specialist producer of chemical reagents for isolation and purification, including affinity chromatography techniques (using antigens/antibody characteristics as a means of specific isolation). It sees itself as performing an essentially 'service role' to firms like Novo which are selling biologically produced products and need help with problems of scale-up, extraction and purification. Its interests in this area have also taken it into the design and manufacture of equipment and instrumentation.

KabiGen is more strictly speaking a 'biotechnology' firm, its main interests being in research into drugs and vaccines which can be produced by rDNA. It also sees its main role as a service one, to 'cell breeders'. KabiVitrum, its parent company, is the world's largest producer of human growth hormone, currently derived from human cadavers, but this source is soon to be replaced by a bio-engineered product which KabiGen has been developing from Genetech's basic 'engineering', and 'scaled up' at the British Porton laboratories (this is an apt illustration of how international the leading-edge technology is). KabiGen's shareholders now include Volvo, Korsnas Marma—the forestry and wood products group—and Alfa Laval. Alfa-Laval have wide interests in equipment for biotechnology and are now the foremost European suppliers.

In many respects both Novo and Pharmacia/Fortia are analogous to some of the American biotechnology companies: small companies which have found a niche in the fast-expanding biotechnology, have cultivated that niche, and prospered. Indeed, it is worth noting that they are the only European stocks to be included in the American stockbroker E. F. Hutton's list of biotechnology growth stocks.

SWITZERLAND

Switzerland is the home of one prominent new biotechnology firm —Biogen—and three major pharmaceutical firms. Biogen is to all intents and purposes an American firm. It was founded by Walter Gilbert, a Nobel prize-winner from Harvard, in 1978, and location in Zurich was partly aimed at attracting funds from European multinationals, partly at using European brains and partly to benefit from Switzerland's favourable tax and disclosure laws. The European multinationals proved less attracted than their American counterparts to investing in brain power. Only one European company, the British Grand Metropolitan, sponsored it, the other three partners being American multinationals (INCO, Monsanto and Schering-Plough). Biogen, however, has established itself as one of the largest and noisiest of the new biotechnology firms. Its interests are firmly in the genetic engineering field and it was the first firm to announce expression of a hepatitis B antigen, leukocyte interferon and a viral antigen for foot and mouth disease. Many of its products are currently undergoing clinical trials, but, like the other major new biotechnoloogy firms, it has yet to market any genetically engineered product. It has been fighting a major patent battle with Genentech over alpha-interferon—the first of many such battles in biotechnology![7]

The three major Swiss pharmaceutical firms—Hoffman La Roche, Ciba-Geigy and Sandoz—are more typically Swiss firms. Ciba-Geigy has the greatest in-house commitment to biotechnology and is currently building a $19.5m biotechnological research centre in Switzerland which will employ over 150 people and work closely with local universities and research institutes. Hoffman La Roche, the largest of the three, already has substantial research facilities in the United States and uses these for its biological research. It has research links with a number of the new biotechnology companies, including Genentech (where links date back to 1978), and has recently concluded an agreement with Takeda Chemical Industries for joint development of interferon. Sandoz also has an American subsidiary and research links with several American partners, giving it access to work on monoclonal antibodies and seed genetics. It has links with University College, London on neurobiological research.

Government-sponsored research in Switzerland is decentralised. The Federal Government supports the Federal Institute for Technology (Eidgenoessische Technische Hochschule) in Zurich which is considered to be one of Europe's leading institutes for biotechnology research. Universities (Zurich, Basle, Berne, Lausanne, Geneva) are funded by the local cantons and support specialist research institutes in molecular biology, genetics, etc. Federal funding is available to help defray the costs, but it is not revealed how much help is given.

The European perspective

For biotechnology, as for other activities, a European perspective must take account of collective EEC thinking, such as it is. It is noteworthy that the earlier discussion in this chapter has needed no reference to the Community, reflecting the fact that until recently there has not even been the beginnings of a European policy. In 1981, however, biotechnology was one of the areas chosen for investigation by FAST (Forecasting Assessment in Science and Technology), the EEC's technology 'think tank'. A recent Commission report on biotechnology based on its work went to the Council of Ministers in June 1983 (European Commission, 1983a). Biotechnology was declared 'a priority area for innovation', and a joint working party, to be chaired by the United Kingdom, has been established to develop a joint Community programme.

The Commission has suggested four objectives for a Community programme:

(i) the need to extend the market for biotechnological products by removing all 'divergent' regulations in the various nations covering foodstuffs and other products;
(ii) the creation of a more effective patent system;
(iii) the development of a joint research and development programme at a Community level;
(iv) the elimination of competitive handicaps resulting from the common agricultural policy—in particular, the need to try to secure access at world market prices for grain, sugar and starch when these are used as a base for biotechnological products.

What has been achieved in relation to these objectives and what is likely to be achieved? Are they sensible objectives in the interests of European biotechnology as a whole, as distinct from the vested interests of the European bureaucracy? Is it correct, as the FAST report suggested, that only a sustained, joint European effort can achieve the 'critical mass needed in modern biotechnology'?

Taking the four objectives, one can note, first, that in all four cases there are strong innate conflicts of interest involved. Of the four objectives, two—the removal of barriers to the internal market and the need to give Community access to cereal and sugar at world market prices—are issues that have arisen many times since the Community's creation in 1957, with little result. One outcome of the Stuttgart summit in 1984 was a Community commitment to reduce the price of cereals used as an industrial feedstock, which is a significant step in the right direction although it remains to be seen whether the dual price system it will necessitate can be effectively policed. It leaves untouched the problem of fructose syrups and the hoary issue of protection to Europe's sugar beet growers —perhaps a salutary reminder that there will always be a problem when a new or alternative foodstuff challenges the market of an established and protected agricultural product. These are the first of many battles that are likely to be fought between the emergent biotechnology industry (which will do so much to industrialise food production) and the farming lobby.

HEALTH AND SAFETY FACTORS

As to the internal market, there is one issue which cries out for effective regulation at the Community level: the issue of the *ethical and safety* aspects of biotechnology. Given the newness of the subject, most countries are still in the process of formulating regulations in this field and there is everything to be said for the new regulations being, from the start, harmonised Community controls rather than the typical pattern of divergent national regulations. There are very real fears, particularly in West Germany, Scandinavia and the Netherlands, that biotechnology is likely to unleash a Pandora's box of mutant strains into the environment. Now that we have several years of experiment and experience to draw upon, there is need for an open and informed debate. This in itself would do much to allay fears, many of which stem from lack of knowledge. There is also a need for standards which are practicable but which reflect the public concern that man, not micro-organism, should be in control.

PATENTS

The other two objectives—a more effective patent system and a Community programme for R&D—raise different issues. On patents, as on regulatory issues, national laws still operate since many member states have still to ratify the Convention on the Community Patent (which would bring into effect a unified Community patent system). Given considerable variation from country to country in the treatment of biotechnology, the need for some unified Community treatment is urgent. But so, too, is the clarification of the ambiguities under the American patent and it is arguable that this, and the alignment of European practice on publication prior to patenting with American practice (which allows a six-month period of grace), would currently be of more value to European biotechnology than protracted negotiations aimed at aligning national practice.

A CO-ORDINATED RESEARCH PROGRAMME

The new issue raised by the Community proposals is that of a co-ordinated research programme. The word co-ordination is perhaps a misnomer. The FAST proposals used the French word 'concertation', and the spirit of the proposals was to create a 'research institute without walls', rather than to co-ordinate and allocate research projects. There is a body of thought in Europe which argues that the Community can only achieve the 'critical mass' necessary for effective R&D in biotechnology by pooling its country-based research programmes into a single Community programme. Such an assumption is not justified. In the United States, although total expenditures are very large, they are dispersed between many institutes, each of which competes with each other, and it is this competition which creates much of the intellectual dynamism.

The concept of 'critical mass' in biotechnology research applies in fact in a somewhat different dimension. What Europe lacks are centres where many different research groups are working in the same broad area and where researchers have the stimulus of frequent contact with others working in similar fields. In other words, the 'critical mass' is in terms of the number of researchers within a geographical area. In the United States there are numerous centres where this is so—Harvard, Philadelphia, Berkeley and Stanford being the obvious examples. In Europe, Cambridge and Heidelberg come to mind: perhaps also Paris and London, but contact between institutes in these capital cities is often not so easy.

If co-ordination therefore means encouraging the growth and interchange of personnel between some of the existing centres of excellence within Europe (and discouraging too much dispersion of activity), then the Commission initiative is to be supported. If it means encouraging more and more frequent contact between research groups, this too is good. But if by co-ordination is meant a grand exercise allocating research priorities between centres and making sure that research programmes do not overlap, then there is very little to be said for it. Given the present state of knowledge, there is much to be said for competition and little to be gained from co-ordination.

Firms, governments and technologies

For both industries and government, biotechnology presents the classic dilemma of the new industrial activity—when to get involved and how deeply to get involved? The scientific and technological base upon which the activity is predicated is itself moving so fast that early involvement risks being overtaken by subsequent developments (or else failure because predictions as to what could be achieved were over-optimistic). But doing nothing risks failure to accumulate sufficient technological expertise to enable entry into the activity at a later point in time. In other words, the dynamic learning curve effects begin at an early point and a firm risks excluding itself at a later stage if it does not invest in expertise. Given the degree of uncertainty which still surrounds the emergence of biotechnology, there is no right balance between these various considerations. The strategies adopted by firms and governments will obviously reflect their own (subjective) judgement of probabilities and attempts to hedge their positions.

FIRM STRATEGIES

As regards firm strategies, the situation varies from sector to sector. Most pharmaceutical companies now view biotechnology not as a revolutionary process but as an evolutionary one. The natural development of medicine towards a greater understanding of the body's own defences against illness led inevitably to an increasing interest in natural anti-viral agents, hormones, etc., and their use in therapeutic medicine. Genetic engineering takes this approach one stage further: it provides a method of manufacturing these natural proteins rather than having to extract them at considerable expense and difficulty from animal or human cell tissue. But this was not always so. In the early days of genetic engineering it was not clear that the technique would succeed in developing such a 'manufacturing' technique, and the emergence of the 'start-up' firms in the United States at this early stage reflected these uncertainties.

The most common observation about the development of biotechnology in Europe relates to the almost total absence of the start-up company. Some see this as in itself an indicator of the relative sluggishness of interest and activity in Europe. This is wrong. Part of the reason lies in Europe's comparatively late take-up of genetic engineering and part lies in the general lack of a cultural and institutional framework in Europe: there is still no developed venture capital market and few governments offer the tax exemptions or stock option relief which play such an important part in encouraging the academic entrepreneur in the United States. Part also lies in the inflexibility of the European academic hierarchies, particularly in France and Germany, which makes it far more difficult for the academic to break out of the mould.

In general, however, European companies, like their Japanese counterparts, have chosen to build up research capacities and knowledge in-house rather than contracting it out. In contrast to Japan, there is relatively little collaborative research in Europe and many companies are highly reticent about their expenditure and their work related to biotechnology.

This secrecy partly reflects the unsatisfactory patent situation. As yet, companies cannot be certain that a genetically engineered organism can be patented. But even where an organism, or the process for obtaining it, can be patented, there remains a question of whether it is better to patent and risk disclosing to competitors work that is being done, or whether it is preferable to keep the work secret, relying on the loyalty of employees. The European pharmaceutical industry has always tended more towards the 'secrecy' route than its American counterparts. The Japanese, in contrast, tend to patent even where they have no intention of further development work.

Outside pharmaceuticals, strategies are far more tentative, reflecting the greater degree of uncertainty. Fine chemicals and commodity chemicals, being the sectors closest to pharmaceuticals, are those that are least dominated by the 'wait and see' syndrome, but they risk being an area that is increasingly crowded as the bulk chemical manufacturers like Bayer, Hoechst and ICI (all of whom have bulk fermentation experience from SCP and sewage techniques, as well as competence in pharmaceuticals) eye the high value added and fast-growing markets. To date, innovation has been incremental rather than discontinuous: chemicals produced by fermentation have benefited from improvements in bioreactor techniques and further improvements are likely to increase yields and extend the range of products manufactured via fermentation. European producers, however, have been slow, compared to the Japanese, in recognising the potential that genetic engineering has for raising yields. The exception here is the Danish firm Novo, but Novo is in all respects an exceptional firm, closer in many senses to the American biotechnology firms than to the traditional European chemical and pharmaceutical giants. (This does, however, raise the question of whether the established oligopolists are creative enough in their thinking and planning to seize the opportunities that a new technology can offer.)

In most of the other areas discussed—bulk chemicals, biomass energy, SCP, food and agriculture and enhanced oil recovery—the application of biotechnology really waits on scientific discovery. For example, genetic engineering in plants will clearly have far-reaching effects on agriculture and food production, but it could be this year, or it could be ten years hence, that the key breakthrough is made. Until such time, firms in the business can do little more than make sure that they are exploiting known technology, such as cloning and plant cell culture techniques, to the full. Many of the larger firms in the food area, for example Unilever and BSN-Danone-Gervais, have built up competence in their own research laboratories and work closely with university and government laboratories. Others, like Grand Metropolitan, have bought a stake in Biogen and set up their own small 'think tank' monitoring organisation. However, it is open to question whether here too European firms are not being ultra-cautious compared with their American or Japanese counterparts. As a recent survey for the *Wall Street Journal* revealed, innovation in Europe is biased towards process innovation, with an extreme conservatism over the development and innovation of new products.[8] Sitting on the fence tends to create a situation in which the sitter is always forced to catch up with someone!

GOVERNMENT STRATEGIES

If firms are faced by uncertainty, so too are governments. Yet all European Governments are under increasing pressure to promote biotechnology as 'an industry of the future'. This pressure arises from two sources, internal and external. The internal pressures derive primarily from the academic establishment (often inter-mixed with the industrial scientific establishment), anxious to ensure for its own disciplines a continuing and preferably increasing slice of the (often diminishing) cake of government funding. The external pressures come from the increasing world paranoia about the Japanese technological challenge. The fact that Japan has declared biotechnology 'an industry of the future' and has committed modest public funding to a co-ordinated research programme brings an immediate demand for equivalent action from all European Governments.

As with the fifth-generation computer, it could be that the real challenge is coming from the United States rather than from the Japanese, for it has yet to be seen whether the Japanese can achieve some of their more ambitious technological targets. In the meantime, all governments are caught up in a technological race that is difficult to abandon and which, in Europe, is additionally complicated by the cross-currents of co-operation versus nationalism. The logical way to meet the Japanese challenge may be to mount a co-ordinated European programme, but there remain doubts as to whether this can be done effectively and, while doubts remain, it becomes a matter of national pride to mount a programme at least equivalent to that of one's European partners.

The figures quoted earlier in this chapter indicate the sheer weight of American spending, particularly on medical research, compared to that of European countries: even more so, if private expenditure on cancer research and related medical disciplines were added to these public expenditures. Biotechnology is in many respects a spin-off from this research spending (just as microelectronics emerged as a spin-off from the space and defence programmes). The United States has no 'strategy' for biotechnology, but it needs no strategy. These vast expenditures on medical and cancer research propel it to the forefront of technology and feed the intellectual dynamism which is such a feature of the American scene.

It is wrong, however, to view this problem in the black and white terms of the technology gap. The very diversity of biotechnology, and the fact that so much waits still for scientific feasibility, means that it is simplistic to present a view of biotechnology, as does the recent OTA Report (Office of Technology Assessment, 1984), in terms of the established and assured pre-eminence of the United States in all aspects of the technology. There is no doubt that the United States currently leads the world in techniques of genetic manipulation but, as was seen earlier in this chapter, there is much to biotechnology that goes beyond genetic manipulation. As *Nature* in its leader on the OTA Report commented, 'The constant harping on the theme of leadership, by representing the process of technological innovation as a battle between two Goliaths, obliterates the texture of what is going on. In such an interesting field there will be opportunities for all and, eventually, a division of labour, Adam Smith style'.[9]

The contrasting strategies adopted by the three major European governments reflect considerable variation in the perception of these opportunities and the risk attached to them. The German Government shared with the Japanese recognition at an early stage of the potential importance of fermentation technology and the need for research and development to go into bioreactor design and downstream processing technologies. The German Government strategy, influenced very substantially by the biotechnology establishment centred on DECHEMA, the chemical plant trade association, has been to build upon German traditional strengths in the chemical sector. This strategy was reinforced by the 1973 energy crisis, the need to develop alternative routes to energy and the strong environmental lobby in Germany. These priorities were rudely shaken in 1981 by the Hoechst decision to go to the United States for its genetic engineering research, and as a result there has been something of a *volte face* on the BMFT programme, with 'catch-up' in genetic engineering and cell fusion topping the list. Nevertheless, considerable emphasis continues on the bioreactor programme.

French strategy illustrates the French willingness to set their sights on a high-risk, high pay-off route with a long-term horizon. Coming late to biotechnology and with relatively little industrial strength behind it, the French Government strategy has been to lead from the front, building on French research strengths in medical and agricultural sciences and, with the exception of the ever-willing Elf-Acquitaine, pushing a somewhat reluctant French industry along with it. Much of the French effort is in catch-up, with a major educational and research initiative in most of the main areas of biotechnology. The two areas which are singled out for special effort are plant genetics, with particular attention to the seed industry, and cellulose degradation, which is seen as linking into the French carburol (biomass energy) programme. Both are areas where research is still very much in the laboratory. It is still uncertain whether the breakthrough will come this year, next year or in ten years time.

In the United Kingdom, particularly in bridging the university/industry gap, the problem has not been 'catch-up' but applications, and Government policy has been largely geared to this end. Unlike Germany, there was no sizeable internal pressure group for biotechnology until fairly late in the 1970s when Government spending cuts on universities mobilised a vocal academic lobby at much the same time as external pressures began to mount. Like the French, Government action has come relatively late in the day but, unlike the French, the British Government has refused to lead from the front and lay down priorities. Indeed, the strategy that developed was more open-ended than either the French or German strategies, putting emphasis above all on collaborative research (university/industry), and leaving it substantially to industry to decide upon priorities. The question, which only time can answer, is whether British industrialists will rise to the challenge. The danger is that in this area, as in others, failure will arise not from lack of competence in the 'leading-edge' technologies, but from an inability to carry this expertise over to more general application. In this respect, Britain's traditional neglect of, and appreciation for, engineering

—in this case chemical engineering and downstream processing—may once again prove an Achilles' heel.

THE SCOPE FOR ACTION

It seems likely that in one form or another the world's needs for food, fuel, clothing and shelter will remain substantially dependent upon the 'bounty of nature'. Biotechnology, in the sense of being a technology which harnesses and improves upon the productivity of nature, is a technology which has been being used and perfected since time immemorial. The 'new biotechnology' discussed in this chapter is both a continuation of this process and a departure from it. Methods of genetic manipulation, already successful and well practised in relation to simple micro-organisms, and potentially applicable to plants and animals, mean that, for the first time, man can go beyond working within the bounds of nature (e.g. by breeding mutant strains) and can manipulate those bounds. This development, as was stressed earlier, is only one of a series of powerful new scientific techniques which together constitute 'the new biotechnology'.

The burden of this chapter has been to stress the wide and diverse impact that the new biotechnology will have upon a wide range of industries. Europe is well placed to take up and profit from this new biotechnology. It has a strong pharmaceutical industry and a chemical industry with traditional strengths in fermentation technology. Its agricultural research base is excellent and it is well to the fore in much of the research that is proceeding on plant genetics. Britain mirrors the United States in being at the leading edge in at least some areas of the medical/pharmaceutical complex: Germany has strength in fermentation technology (which to a degree the United States lacks, but is Japan's great strength): France and the Netherlands have the agricultural expertise: Novo Industri are already world leaders in enzyme technology. Left to itself, there can be little doubt that biotechnology would take root and flourish in Europe: there would be some areas where Europe would develop comparative advantage in both technology and production and others where it would cede such advantage to other countries.

To suggest, however, that biotechnology should be 'left to itself' is unrealistic, for it has already been caught up in the high technology 'race' which has developed between the advanced industrial countries, in particular between Japan and the United States. For Europe to opt out of the race is seen as leaving the field to the United States and Japan and accepting that in this area it should fall within the technological hegemony of one or other of these two countries. Unfortunately, such are national sensibilities, that 'opting in' adds not one, but at least three contenders to the field.

The question therefore becomes: 'How may Europe, either as the EEC or as individual governments, best promote the development of biotechnology within member states?' Although there is much talk of 'co-ordination' at a European level, there is no case for a grand European strategy. Unlike nuclear fusion, research in this area does not require major capital expenditures. Much of it can be conducted within the confines of well-equipped university laboratories, the costs of which are well within the bounds of national R&D budgets. Even pilot fermentation facilities, though not inexpensive, are nevertheless but a fraction of the cost of the JET or CERN nuclear projects.

There is much therefore to be said for a decentralised rather than a centralised approach. This conclusion is further reinforced by the uncertainty which dominates so much of the area. The new biotechnology is still young: hopes abound, but as yet cloned interferon remains a chemical in search of a disease to cure, just as genetically-manipulated plants that produce their own fertiliser are a splendid idea but cannot yet be bred. Neither firms, Governments, nor the Community can predict which research is likely to yield the greatest pay-off short-term, or long-term. In such circumstances, there is much to be said for competition—both between research establishments and between firms.

Given this decentralised approach, the role for governments and the Community becomes a matter of 'promoting the right environment'. The following provides a broad check-list of policies:

(i) *Support basic research*: time and again it is apparent that benefit derives to those countries which have a broadly-based research effort in the basic sciences underlying biotechnology and a strong cadre of teachers and researchers.
(ii) *Encourage technology transfer between academic science and industry*: through, on the one hand, collaborative programmes and, on the other, interchange of personnel and ideas. This may be achieved via the market, as in the United States where the new biotechnology companies act as an intermediary between academic science and industry; but Europe currently lacks the market institutions (and mentality) of the United States and the collaborative route for the present may be preferable. For this to work well in an interdisciplinary field such as biotechnology, where cross-fertilisation of ideas is so important, there

must be open and free exchange of information and knowledge. This may require a generally acknowledged code of practice on the appropriate balance between commercial secrecy and dissemination of results.
(iii) *Encourage the development of a venture capital market*: through tax concessions, stock options, etc. Europe may not have the American pioneer mentality, but institutions do help to create an environment in which new firms can flourish, bring new products on to the market and challenge the complacency of existing firms. Such firms can also provide a route to technology transfer. The example of Celltech in the United Kingdom illustrates how Governments may make creative use of the venture capital market *'pour encourager les autres'*.
(iv) *Encourage the European initiatives*: on health and safety regulations, patents and cereal prices. Support concertation in the form of information and personnel exchange, but be wary of grand European strategies aimed at avoiding duplication in research (for this cannot be justified on grounds of economies of scale).
(v) *Encourage joint ventures and licensing*: biotechnology is too broad and diverse a field for any one country to dominate the leading edge in *all* areas of activity. Two lessons are to be learned from the success of countries such as Sweden, Denmark and Switzerland. The first is that benefit comes from applying a new technology rather than necessarily from participating in its discovery and development. Second, that maximum benefit comes from selective use. A high level of competence can be built up and exploited in particular areas by not being ashamed to buy in expertise, marrying it up with internal capabilities.
(vi) *Make creative use of public purchasing*: in most of Europe, governments play a considerable role in the provision of health services, yet it has been left to the pharmaceutical companies to decide on priorities for research: similarly, there is considerable scope in the treatment of sewage and other waste materials for Governments both to improve processing and promote biotechnology. Some of the other studies in this book (e.g. offshore supplies) indicate how a close customer/contractors relationship in the early stages of development can help a new activity to establish itself strongly in world markets.

This decentralised approach to policy has the additional advantage of emphasising the importance of decision-taking by the individual firm. The focus of this study is on *how* biotechnology is being translated from research laboratory to firm. There is little doubt that the key factor in this process is the firm and that the vital decision as to whether a new idea is worth commercialising and how it shall be done is best left to the firm (and/or its financial advisers). In Europe, the decision-makers that count are the major firms in the chemical, pharmaceutical and agri-food business: already Hoechst and Bayer are each committing resources to research in biotechnology roughly twice those of the Federal Government. Governments may seek to influence, persuade or jockey these firms into taking different decisions, but short of a major public enterprise initiative (which runs into all the problems of uncertainty discussed above), it is their decision, not Government decisions, that count.

The problem for European Governments is the danger that too many of these enterprises, having established in-house capabilities, are now sitting back, watching and waiting for others to move. In the United States, the competitive spur provided by the small firm sector (and the transparency it provides on research contracts) has forced the major multinationals off the fence: in Japan, the mutual support provided by collaborative research similarly helps to promote commitment. What Europe needs perhaps more than anything else at present is leadership which demonstrates a long-term confidence in this still very young, but potentially very important, technology. As a recent leader on the subject in *Nature* remarked: 'short-term uncertainty against a background of long-term confidence should be the motto for the future.'[10]

Notes

1. The discovery of restriction enzymes is a classic case of the vital breakthrough coming from the pursuit of pure science. Herbert Boyer had set himself the task of identifying and classifying all enzymes found within living cells—a task very much in the traditions of classic microbiology. In the process he discovered the category of enzymes now known as restriction enzymes and quickly recognised their function as of major significance.
2. In fact the first genetically engineered product to be sold on the open market was a scours vaccine (for pigs), developed and marketed by Akzo (Intravet) in the Netherlands.
3. In January 1983, 196 Japanese, including fifteen from leading industrial countries, were working at the National Institutes of Health in Washington DC. This was double the number of any other country, India, with 97, being the next highest, followed by Italy (91), the United Kingdom (69) and France (52). (See US Government Interagency Report, 1983, p. B.76a.)
4. For details, see *Business Week*, 14 December 1981, pp. 73–4.

5. This view was expressed in interview with a Hoechst executive. The deal involves Hoechst in financing over a ten-year period, at $6m per annum, a research institute under Dr Howard Goodman at the MGH, which in turn has research links with the Harvard Medical School and MIT. Under the agreement, Hoechst has first option to exploit new syntheses developed within the research institute (although the US Federal Government has first option rights on any development funded primarily with US Government monies) and also has its research workers working (and training) alongside the Americans. From Hoechst's point of view, the gain is as much in bringing the 'lateral thinking' of the American scientist to bear on the whole range of Hoechst production, as well as in terms of specific genetically engineered products which may emerge from the 'affair'.
6. Celltech was in fact set up both to exploit the potential of monoclonal antibodies which had been pioneered in the Cambridge MRC molecular biology laboratory, and to prevent a repeat of the fiasco which resulted in monoclonal antibodies not being patented. Initially, Celltech had first option rights to all genetic engineering developments in MRC laboratories. Subsequently, this was modified. Celltech now has very limited first option rights. See Sharp (1985), Chap. 7, footnotes 52 and 53.
7. In fact the battle was between Schering-Plough, which acquired the rights to Brogen's alpha-interferon in 1979, and Hoffman La Roche which has the rights to Genentech's product. The legal battle was over the issue of which company rightfully owned the technique for mass producing alpha-interferon via fermentation, a tricky issue since Genentech was the first to isolate the alpha-interferon gene, but Biogen was the first to learn how to implant the gene into mass-producing bacteria. Agreement has eventually been reached (May 1985) effectively dividing the market between the two groupings, with Biogen/Schering-Plough concentrating on Europe, and Genentech/Roche on the US markets. Each will also market their rival's product. In the meantime, Biogen has been in considerable trouble with the resignation first of its chariman, and founder, Dr Walter Gilbect in December 1984, and subsequently of Dr Julian Davies, director of its Geneva operations. Its plight is typical of that of new biotechnology firms, particularly those which have concentrated on developing the therapeutic proteins via genetic engineering (rather than diagnostic products). Since 1980 it has spent $217m, but cash income is neglible, implying a continuing need to borrow in an increasingly difficult market. Cash difficulties may well have influenced the Schering-Plough/Brogen decision not to pursue the patent battle through ever more costly litigation. See S. Yanchinski, 'US giants squabble over drug patent', *Financial Times*, 12 April 1985, and J. Erlichman, 'Drug giants make interferon truce', *The Guardian*, 22 May 1985.
8. See *Wall Street Journal*, 31 January 1984. Special survey on new technologies in Europe.
9. *Nature*, 2 February 1984, p. 399.
10. Ibid.

Bibliography

ACARD (Advisory Council for Applied Research and Development) (1980), 'Biotechnology: Report of Joint Working Party of ACARD and Advisory Board for the Research Councils and Royal Society'. (*The Spinks Report*), London, HMSO.

Apeldoorn, J. H. (1981), *Biotechnology: A Dutch Perspective*, STT Publications No. 30, Delft University Press.

Brech, M. and Sharp, M. (1984), *Inward Investment—Policy Options for the United Kingdom*, Chatham House Paper No. 21, London, Routledge and Kegan Paul.

Cantley, M. F. (1983). *Plan by Objective: Biotechnology*, European Commission FAST Project XII/37/83/EN, Brussels, Commission of the European Communities.

Davies, J. (1983), 'Watchful eye being kept on biotechnology', *The Financial Times*, supplement on West Germany, London, 4 May.

Deutsche Gesellschaft für Chemisches Apparatwesen (DECHEMA) (1974), 'Biotechnologie', Report published for BMFT.

Dunnett, J. (1983), 'Time to Count the Costs', *Nature*, 301, 3 February, p. 367.

Dunnill, P. and Rudd, M. (1984), 'Biotechnology and British Industry'. A report for the Biotechnology Directorate of the Science and Engineering Research Council, Swindon, SERC, UK.

The Economist (1981), 'Kill or cure for Japan's drug firms', 29 August, pp. 78-9.

The Economist (1982), 'Japanese drug companies—euthanasia', 11 December, p. 78.

The Economist (1984), 'Japan cannot find the right pick-me-up for its pill makers', 16 March, p. 63.

European Commission (1983a), 'Biotechnology: The Community's Role'. Background Note on National Initiatives for the Support of Biotechnology, Brussels, European Commission, Com (83), 328, Final 2.

European Commission (1983b), *Forecasting and Assessment in Science and Technology (FAST) Report*, Vol. 1—Results and Recommendations; Vol. 2—Summaries of 36 research projects, Brussels, Commission of the European Communities.

Eurostat (1984), 'Public Expenditures on Research and Development, 1975-82'. (Includes special report on biotechnology.) Brussels, Statistical Office of the European Communities.

Gros, F., Jacob, F. and Royer, P. (1979), *Sciences de la Vie et Société*, La Documentation française.

Howells, E. R. (1982), 'Single Cell Protein and Related Technology', *Chemistry and Industry*, August.

Japanese External Trade Research Organisation (JETRO) (1982), 'Special Report on Research in Biotechnology in Japan, published by JETRO, in English.

Junne, G. (1984), *Biotechnologie: Das Gesetz vom bremsenden Vorsprung*, Chap. 1 in Peter H. Metler (ed.), *Wohin expandieren Multinationale Conceonen?*, Frankfurt, Haag and Herschen.

Milanesi, G. (1984), *Le biotechnologie in Italia*, Italy, Federation of Scientific and Technical Associations (FSTA).
New Scientist (1983), 'Plant genetics: new frontiers', 5 February, p. 87.
Office of Technology Assessment (OTA) (1982), *Genetic Technology: A New Frontier*, Westview Press/Croom Helm.
Office of Technology Assessment (OTA) (1984), *Commercial Biotechnology— An International Analysis*, Washington, DC, Government Printing Office.
Pelissolo, J. (1980), *La Biotechnologie Demain*, La Documentation française.
Rogers, M. D. (1982), 'The Japanese Government's Role in Biotechnology', *Chemistry and Industry*, 7 August 1982, pp. 533–7.
de Rosnay, J. (1979), *Biotéchnologie et Bioindustrie*, Paris, La Documentation française.
Sargeant, K. (1984), *Biotechnology: Connectedness and Dematerialisation: The Strategic Challenge to Europe*. Paper presented to Biotechnology '84, 1–2, organised by Royal Irish Academy and Society for General Microbiology.
Schilperoort, R. A. (1982), *Innovatie Programma Biotechnologie*, Voorlichtingsdienst, Wietenschapsbelerd, The Hague.
Sharp, M. (1985), 'The New Biotechnology: European Governments in search of a strategy', Sussex European Paper No. 15, University of Sussex.
Spinks Report—see ACARD (1980).
Surrey, J. and Walker, W. (1981), 'The European Power Plant Industry: Structural responses to international market pressures', Sussex European Papers No. 12, University of Sussex.
UK Government (1981), 'Biotechnology', Cmnd 8177, March.
US Government (1983), *Interagency Working Group on Competitive and Transfer Aspects of Biotechnology*, published by Elsevier Science Publishers, Amsterdam, for McGraw-Hill's *Biotechnology Newswatch*.
Yanchinski, S. (1982), 'France sets its sights on biotechnology future', *The Financial Times*, 3 August.

7 The offshore supplies industry: fast, continuous and incremental change

P. Lesley Cook

Newness is a matter of degree

The Offshore Supplies Industry is the general term used for the great collection of firms which supplies hardware and services to the oil companies which explore, develop and produce offshore oil and gas.

It is a new industry in several senses. First, it is relatively new world-wide. Shallow water drilling began before 1939, notably in Lake Maracaibo in Venezuela, but it was not until after 1945 that offshore drilling became significant in the Gulf of Mexico, and by the 1950s there were also developments in the Middle East, the Mediterranean and off West Africa. In all cases, however, the water was quite shallow—under 100 ft. deep—and relatively protected.

American firms have dominated the industry. All the big offshore firms are part of groups which are major suppliers of equipment and services to the oil industry. They cut their teeth on the large American market and went overseas with the multinational oil companies; one important result is the widespread use of American standards and specifications. The major American offshore firms were established in the early 1950s; and several of the firms involved, notably McDermott and Brown and Root, rapidly became world leaders. These firms were employed, often on turnkey contracts, to build the relatively small platforms then required.

Of European countries, both France and Italy have built up a degree of expertise. The Italians had been much involved in the major pipeline projects required to bring gas from Algeria and Libya across the Mediterranean; the French national oil companies were seeking oil supplies and were exploring actively in West Africa. In Italy and France work was done by nationalised firms employing nationalised or semi-nationalised contractors and domestic subcontractors. The French and the Italians were also involved in the Middle East, as was BP; much of the early British experience was obtained by firms working with BP off Abu Dhabi.

The second sense in which the industry is new is that it is new to Europe. In 1959 the Dutch found gas off their coast; by the early 1960s there was intense surveying work in the British sector of the North Sea. Drilling began after the passing of the Continental Shelf Act in 1964 and gas was found in the southern basin in December 1965. Development proceeded rapidly in water depths of 50 to 100 ft., comparable with those in the Gulf of Mexico, and the platforms required were similar to those used in the Gulf;[1] indeed, in all respects experience gained in the Gulf of Mexico was wholly relevant (which gave the major American firms an immediate advantage) except that the North Sea weather was much more severe. The experience gained from the early contracts, of coping with the more difficult weather conditions, was of considerable importance in subsequent exploration and development.

Drilling in the relatively deep waters of the northern basin of the North Sea began in the late 1960s and oil was found in the Norwegian sector in 1969. Other discoveries followed rapidly. By 1969, it was clear that major investment would take place and by 1972 the orders were being placed. In 1973 the price of oil quadrupled and with it the profitability of the North Sea. As a result, development became urgent and the new industry expanded very rapidly.

Newness is very much a matter of degree and it is misleading to draw a hard line between offshore and onshore oil production as they have very much in common. The similarities are particularly close in drilling technology and equipment, well-logging, reservoir assessment and the control of wells. (See Glossary at the end of this chapter for technical terms.) The oil companies, the drilling companies, and the equipment manufacturers all had much of the basic expertise required for offshore work. Nevertheless, much of the equipment and the techniques have been refined and adapted in response to the offshore environment and, particularly, to the very bad weather conditions. Both make the cost of exploration wells much higher.

It is incorrect, therefore, to think of the discovery of oil in the North Sea as creating a demand for completely new products and services. The severity of the conditions and the enormous problems of scaling up designs and coping with new problems did mean quite a jump to meeting demands which were, however valuable previous experience, in some degree new. Meeting these demands resulted in a continuous learning process—those entering the industry entered a dynamic and continuously changing activity.

It is also important to recognise that, while the actual products and services required were new to Europe, many of the activities were very similar to those being carried out by existing firms for other purposes. The fabrication of platforms required heavy structural steelwork. The building of oil and gas processing units, the manufacture of power plant and pumps and compressors, the building and running of ships and the survey work, were all activities which already existed. Some firms had to make major changes, while for others the adjustment was relatively minor. But change, where it was required, was often required quickly. Firms had to be able to adapt themselves, their staffs and their techniques to new design standards, new time schedules and extensive design changes; indeed, one of the particularly interesting aspects of this study is the competitive importance for existing firms of this adaptability.

The main characteristics of the new activity

There are a number of important characteristics of the industry which are worth highlighting.

(i) DIVERSITY

The goods and services required and the types of firm supplying them are extraordinarily heterogeneous. Small firms, large firms, multinationals, diversified firms and specialist firms were all involved. Requirements in terms of know-how and capital differ appreciably between activities for within each activity many different types of firm are operating side by side. Geophysical surveying, exploration and appraisal drilling, the design and fabrication of production platforms and drilling rigs, pipe laying and platform installation, diving and underwater vehicle design and operation, specialist ships, helicopter services and other support services, could all be regarded as separate 'industries'. The diversity is such that few policies will affect all sectors and most will be effective over a relatively small range of the industry.

(ii) LARGE AND COMPLEX PROJECTS

The offshore supplies industry is best seen as a large sub-contracting network. Each major project is very large and will involve numerous firms as sub-contractors. For example, each large production platform and associated developments in the northern North Sea is likely to cost over £1,000m (1984 prices), and even a new semi-submersible drilling rig will cost over £50m. The number of projects in the British sector is relatively small—five or six new projects a year is high.

The large size of the projects and the high cost of delays means that even the very large contractors are unable to give guarantees which are commensurate with the cost to the client of lateness or unreliability. It follows that the risks of failure by contractors and sub-contractors are effectively borne by the oil company clients and this has led to their increased involvement in the management of projects.[2] In particular, they control the allocation not only of the main contracts but also of the major sub-contracts. Further, they have developed the expertise to take the main technological decisions.

Broadly, the projects involve two types of problems. On the one hand there are the technological decisions and their associated risks. These are largely associated with the problems of scaling up (e.g. much larger platforms and larger modules) and those associated with new concepts such as integrated decks, wholly new types of jackets such as concrete platforms, or, more dramatically, the tension-leg platform (which depends upon a buoyant structure as opposed to a rigid structure).

The other type of problem is managerial and relates to logistics, procurement and quality. These are made particularly difficult by the combined effect of great urgency and multiple design changes (some of which are related to recent experience on other platofrms and some of which stem from further investigation of the oil reservoir itself). The large offshore developments are more difficult to manage than onshore construction because delayed completion of one section holds up the project very seriously; the work must be co-ordinated very precisely. Further, weight is important, as any excess weight may make major modification necessary. Managerial reliability and proven components are thus prime factors in the choice of contractors and sub-contractors.

Jacket design, fabrication and delivery to offshore location involves novel rather than very high technology engineering. Exploration and drilling involves high technical sophistication and skill—the high cost of offshore operations, the importance of correct location and correct platform facilities make this work critical and the oil companies are deeply involved.

The sheer size of the North Sea projects has also led to greater involvement by the oil company client in the whole operation. They have built up in-house competence and moved from prime contractors towards mixed teams and specific project management. Increasingly, these involve a number of major contracting and management engineering firms.

The total number of firms involved in projects is high. There are the engineering consultants and major contractors, who themselves sub-contract some of the engineering work, and more specialised firms for particular jobs. Most of these firms sub-contract further. 'Big fleas have little fleas upon their backs to bite them and little fleas have smaller fleas and so *ad infinitum*.'

The oil company in charge of the development (which is usually owned by a consortium) and the major contractors control the choice of sub-contractors and are therefore a pivotal point in the industry, in that they determine the allocation of work and hence the orders obtained by particular firms or countries. However, this should not be exaggerated; it is almost equally true that it is only when sub-contractors are efficient that major contractors can obtain work. In any country, success depends on both its major contractors and the sub-contractors.

(iii) DEMAND FLUCTUATIONS

Pronounced fluctuations in demand and workload are endemic in the activities relating to exploration and large construction projects. This in part explains the need for an industrial structure which can cope with varying levels of activity. Sub-contracting is a means of risk-spreading since many of the firms doing sub-contracts are diversified. Another important response to demand fluctuations is the maintenance of great flexibility in the size of the labour force; this is achieved by the extensive use of temporary labour provided by specialist employment agencies. Another response is to seek to work internationally and to try to offset reduced demand in one area by moving to another. This applies particularly to drilling, heavy lift installation and pipelaying. In all these activities hire rates fluctuate considerably and the risks are high.

(iv) TECHNICAL CHANGE

Technical progress in the offshore industry has been, and continues to be, very rapid. Not only did work in the North Sea present new problems but movement to more distant waters and the exploitation of smaller reserves have kept up the pressure for new solutions. In addition, the offshore business has many uses for modern electronics (e.g. instruments and control systems) and here new techniques are advancing rapidly. Some of the changes have been quite dramatic.

The early platforms were steel structures but by the early 1970s large concrete platforms which could be floated out with the deck structure complete were expected to reduce hook-up and fitting costs, and to reduce maintenance costs. However, four developments changed the situation. First, heavy lifting capabilities were raised from under 1,000 tons to over 3,000 tons in less than ten years. Second, semi-submersible heavy lifting barges increased the 'weather window' because they could work in more severe conditions. Third, improved hammers (also mounted on semi-submersibles) and piling techniques reduce the cost of piling the foundation of steel platforms. Fourth, improvements in the design of the decks of steel platforms and the greater integration of the deck structure with the modules made platforms more compact and reduced the amount of steel and hence the weight required. These advances meant that by the later 1970s steel was back as the preferred construction material.

Another example of rapid change is in underwater work. In the 1960s, diving was limited to about 100 ft.; it is now possible to go down to 1,000 ft. Further, diving has in part been replaced by use of small one- or two-man submersibles and remotely operated vehicles and this again has made underwater wellheads and underwater completion systems more attractive. The combined effects of progress in underwater systems, flexible hoses, anchoring systems and dynamic positioning has increased the scope for floating systems as opposed to fixed platforms for the exploration of smaller oil reserves.

The really significant advances are matched in importance by the very large numbers of improvements; there has been a continucus and important learning process. One of the most important is the major advance in metallurgy and welding technology. In another area—surveying—modern instruments have made the business highly sophisticated and very much more effective; ten years ago the surveying business was comparatively simple but firms which failed to keep pace with new developments have collapsed and new firms have emerged. A survey ship costing as little as £100,000 may now carry £500,000 worth of equipment.

(v) THE NATURE OF RESEARCH AND DEVELOPMENT

The effectiveness of policies for promoting technical change depends very much upon the nature of the research and development activities in an industry. In the offshore industry these are, as one might expect, varied. First, there is a certain amount of basic research which can range from oceanography and marine geology to electronics (necessary, for example, for the development of new instrumentation for measurement, etc.). This basic research is normally done in research laboratories sponsored either by the Government or co-operatively by the industry. Second, there is the application of the results of this kind of work. This may be called applied basic research and is very often done on a commercial basis, either by one of the research laboratories or sometimes by small 'contract research' firms for a particular client. Third, there is long-range speculative concept development work—for example, new types of platforms and jackets and remotely-operated sub-sea vehicles (ROVs). This is done partly by individual oil companies and partly by consortia. It is of great importance and is basic to the development of new ways of doing things. It is also important to the firms which become involved in this work as they are then on the inside track for subsequent work, should the concept prove successful. Fourth, there is the type of development work which can be done by small firms and is very often done by new firms which have a particular expertise; work of this type has been particularly important in surveying and underwater work. Lastly, a good many major developments come from the interaction between users and the equipment manufacturers and much of the incremental change is dependent upon close user-producer relationships.

As we shall see, the wide variety of R & D work requires a range of policies tailored to specific needs. What is most important, however, is that with the exception of the first category—basic research—R&D in this industry is very much a matter of 'learning by doing'. It is not the type of disembodied R&D that economists frequently envisage. This is the main reason why technical change tends to be continuous and incremental, and why those firms with experience in the industry and large enough to be involved in this continuous R&D process tend to have an advantage over outsiders. It also helps to explain the continuing pre-eminence of the major American contracting groups.

(vi) AN INTERNATIONAL INDUSTRY

The major oil companies operate internationally, as do a large number of the leading supply firms. Although the North Sea exploration and development presented special problems, there has been offshore development in the United States and also an increasing amount of offshore work in many other areas, for example, Brazil, West Africa, South-East Asia, China and Australia. The total world market for offshore products and services in 1984 is estimated at $38bn at 1983 prices (see Scottish Development Agency, 1985); the total British market in 1984 was £9,904m ($13,384m) (NEDO,

1985, p. 6). The North Sea is the largest single market and the technical problems are in general the most difficult, but even the European market as a whole represents only some 25 per cent of the total world market. The international market available to exporters is very much smaller than the total market. First, some of the work must be done on location; second, many of the countries with offshore oil wish to secure a high proportion of the work for domestic firms; and third, many countries seek joint ventures and both a degree of control and technology transfer. The proportion of markets open to exports tends to diminish; whereas five years ago the Far East market was very open, it is now fairly restricted and joint venture firms are increasingly necessary. It is estimated that approximately one-third of the world market is available to exports (SDA op. cit.), but this may well be somewhat optimistically high.

Thus, international competition is mainly in the more technically and managerially demanding activities and increasingly it involves the long-term strategic planning required for joint ventures. The main international markets are in contracting engineering and installation, seismic surveying, exploration drilling, well-logging and drilling and well-head equipment. The leading (mainly American) firms are particularly strong in these areas.

Success in the international markets depends to a large extent upon efficiency, reliability and the ability to solve problems. A good reputation with the major oil company clients is of paramount importance as the selection of major contractors and suppliers lies with them. However, as they are advised by their major contractors and as those contractors select (perhaps subject to monitoring by the client who has a veto) the main subcontractors, their experience of working with the sub-contractors is also important. Links with leading international companies and contractors accounts for much; the oil industry is less exclusively American than it was but they are still fairly dominant. Should either the French or the Japanese, both of whom have tended to purchase nationalistically, become operators for a significant proportion of the new developments world-wide, opportunities for others would be restricted.

Country-based experience

THE UNITED KINGDOM

Experience up to 1973
British firms played a quite subsidiary part in the exploration and development of the gasfields in the southern basin of the North Sea.

Work was largely undertaken by major American companies which did most of the exploration and drilling, and virtually all the project management (Brown and Root, and McDermott); the drilling equipment was largely American. The fabrication of the platforms was done in Europe, with Holland being the most important source. The most significant contribution in the early period by British firms was the building of twelve drill-rigs between 1959 and 1966; the experience was expensive and only one firm, John Brown, was still persevering by 1972 (IMEG, 1972, p. 89).

It was widely assumed, particularly following the 1967 devaluation, that British industry was, or was about to be, competitive. The failure to use this early period to gain valuable experience was partly due to the widely-held view that the market was transitory (Smith, 1978). Certainly, few believed that even if offshore oil was found at depths of over 300 ft. it would be possible to produce it at a cost which could compete with the Middle East oil which was then selling at around $1 a barrel.

Expectations changed sharply in December 1969 with the discovery of the large Ekofisk field on the Norwegian side of the median line. The price of oil was rising and there was a great desire to limit dependence on the Middle East. By 1971/2 large contracts for the first platforms were being placed and questions began to be asked about the low share of the work being obtained by British industry. The situation was sharpened both by what was then seen as heavy unemployment in Scotland and the north of England, and by the Scottish Nationalists' vociferous claim for independence and ownership of 'their' oil.

In 1972 the Department of Trade and Industry commissioned a report from a consulting firm, IMEG—International Management and Engineering Group of Great Britain Limited (IMEG, 1972). If the commissioning of the report represented anxiety, the report itself certainly makes sober reading. British industry was obtaining only about 25 per cent of the contracts and no British firms had achieved significance in any of the main sectors of the industry. The IMEG group drew a sharp distinction between British firms and American firms operating subsidiaries in the United Kingdom. These were of increasing importance. Poor British performance was underlined by the achievements of the French firms, notably Forex (exploration), Comex (underwater), EPTM (pipe-laying), GC Doris (platforms), CFEM (rigs and fabrication) and UIE (fabrication and design). The Dutch had designed and built rigs, drillships, modules and platforms. The Norwegians had been particularly successful in the building of drilling rigs. Vickers Oceanonics who worked in the design and operation of sea-bed vehicles was claimed as a shining example of British achievement (the fragility of that situation was soon apparent as the company pulled out in 1977).

Experience since 1973

The year 1973 was a landmark year. By the end of 1973 the price of oil had quadrupled and the security of Middle East supplies had become even more doubtful. The North Sea discoveries were clearly going to be highly profitable and the race was on. Both Government and the oil companies wanted production as soon as possible—the government for balance of payments reasons and the oil companies because delay was expensive in terms of profits lost.

Demand built up very quickly and the level of orders placed reached a peak by 1974; work completed rose somewhat more slowly but reached a peak (in real terms) in 1976. Table 7.1 illustrates the build-up of work.

Table 7.1 also demonstrates British shares in total contracts. Three-quarters of the overall British share in 1977–81 derived from construction-related development work. As the age of offshore structures increased, there was growing demand for maintenance and underwater services and in this category the British share in 1979/82 was 70 per cent. However, in exploration activity the British share showed no substantial improvement over the 33 per cent obtained in 1974. This is further elaborated in Table 7.2.

Table 7.1 Offshore supply contracts in British sector of North Sea, 1974–1980*

	Value of orders placed, £m (UK share %)			
	Total	Exploration	Development	Maintenance & support services
1974	1279 (40)	180 (33)	996 (40)	103 (58)
1975	1185 (52)	208 (39)	768 (52)	209 (62)
1976	1041 (57)	149 (36)	677 (58)	215 (66)
1977	1295 (62)	75 (33)	829 (69)	391 (53)
1978	1574 (66)	83 (27)	984 (70)	507 (64)
1979	2679 (79)	77 (40)	2042 (83)	560 (70)
1980	2380 (71)	254 (34)	1599 (78)	527 (65)
1981	2911 (67)	552 (32)	1583 (75)	776 (74)
1982	2264 (73)	248 (37)	1087 (80)	929 (73)
1983	2610 (72)	494 (37)	1226 (83)	890 (76)

* At current prices; note slight change in compilation after 1976.
Source: Department of Energy: Blue Books and Brown Books.

British shares of steel platforms and module fabrication, manufactured plant and equipment for platforms, pipe accessories (coating, fittings and castings), maintenance and helicopter services for transporting offshore crews, were high from the outset. By contrast, British shares have remained low in activities requiring offshore-specific facilities and know-how, i.e. exploration drilling, heavy lift installation work and pipe-laying. The British share of drilling equipment is relatively high but this comes very largely from American subsidiaries established in the United Kingdom.

The development of industry in Britain over the last fifteen years has a number of features which are worth noting.

(i) London has retained its position as the main design, engineering and project management and purchasing centre in Europe. The majority of the international firms have branches there and many British firms are major sub-contractors. The oil industry, like the process plant industry, purchases internationally, but the fact that London is a major centre has enabled many British sub-contractors and suppliers of equipment to obtain orders. Over the last fifteen years, the dominance of American designs and American equipment has diminished.

(ii) Although a number of large platform yards were created by large contractors (many with foreign partners), the majority experienced great difficulties. The two most successful were those belonging to McDermott and to Brown and Root/Wimpey.

(iii) Modules have been built by a considerable number of British firms, but again great difficulties were experienced and some of the early ones have gone out and others are not now important. Late entry appears to have been no disadvantage and many of the leading firms today entered the business seriously at quite a late stage. New firms have come in as recently as 1977. Many of the fabrication yards are now partially or completely foreign owned; in addition to Brown and Root and McDermott UIE, Heerema, Grottint and Gotawerk Arendal all have fabrication yards in the United Kingdom. It would not have been possible to pick a winner in this area.

(iv) British firms have done increasingly well in the activities in which they already had good capabilities and product ranges which required only moderate adaptation to meet offshore requirements. GEC Marconi, Ferranti and Plessey have received virtually all the orders for communication systems and gas turbines have been supplied by GEC, Rolls Royce and John Brown.

(v) British firms have done badly in the areas where the existing

Table 7.2 British shares for each offshore supplies activity, 1974–1983

	Percentage of total orders placed	UK share of value of orders placed, per cent*									
	1979–82	1974	1975	1976	1977	1978	1979	1980	1981	1982	1983
Exploration											
Surveys	1	79	62	64	67	73	79	75	47	73	81
Rig hire		25	33	26							
Specialist materials		62	56	57							
Exploration & appraisal drilling	10				26	19	29	31	31	29	34
Development											
Production platforms	27	46	75	76	63	74	84	77	77	88	91
Modules & other fabrications		50	60	53							
Ancillary plant & equipment		75	83	82							
Plant & equipment	8				72	74	73	76	78	86	84
Installation operations	7	6	19	34	66	41	70	54	66	65	66
Development drilling	4				35	70	69	57	71	75	64
Pipe		7	5	33							
Pipe accessories		62	68	86							
Pipelaying		18	11	27							
Submarine pipelines	5				16	38	35	40	56	61	61
Terminals	9				96	85	98	99	98	96	99
Engineering, design & consultancy		60	60	82							

The offshore supplies industry 225

Maintenance & support services	3				33	73	91	86	88	88	91
Inspection testing & maintenance		69	69	67							
Helicopter services		60	59	60							
Marine transport		55	42	40							
Transport	8				50	84	74	61	73	74	73
Diving & underwater services	3	42	69	61	49	77	62	54	81	75	85
Drilling tools & equipment	8	47	52	54	57	52	62	58	69	68	69
Support for personnel offshore	3				44	30	36	39	51	50	57
Other support services	4	68	89	87	81	33	74	98	92	96	95
TOTAL	100	40	52	57	62	66	79	71	67	73	72

*Note difference in classification between the years before and including 1976, and subsequent years.

Source: Department of Energy: Brown and Blue books.

firms were already tending to lose market shares. The prime examples are shipbuilding and shipping.[3]

(vi) Drilling equipment has been supplied almost exclusively by American firms. Much of it is manufactured by wholly-owned subsidiaries in the United Kingdom. It often incorporates major components from the United States. Some specialised or new equipment is imported from the United States or United States subsidiaries on the continent (for example blow-out preventers from France).

(vii) A considerable number of British firms have come into survey work and underwater engineering and some are now working world-wide and competing successfully in areas of high technology; none are large and none are leaders over a wide range of the activity.

(viii) No major heavy-lift barges or pipe-lay barges are owned or operated by British firms. Both activities are large-scale, capital-intensive and risky and they require a high level of expertise and know-how.

Three generalisations emerge from the previous section. First, large British firms have been successful in activities where they already had the relevant expertise and were seen as successful firms. Second, many of the large firms which decided, after oil had been discovered, to try to enter on a large scale have been relatively unsuccessful and it is the foreign subsidiaries or joint ventures with oil-related experience which have become increasingly important. Third, small innovative and flexible firms have been able to grow and compete successfully in niches in the market. The ability to do this has been increased with the growing availability in the United Kingdom of people who, by working in association with American companies, have gained the necessary experience in the offshore industry.

Government policy
Until the 1970s, British Government policy was almost completely *laissez-faire*. The Conservative Government of 1970-4 responded to the increasing anxiety about the apparent failure of British firms to play a large part in offshore development by setting up the Offshore Supplies Office (OSO). The prime business of the OSO was to try to ensure that British firms had a fair crack of the whip and were given a chance to enter the industry ('full and fair opportunity'). Its role was essentially that of monitoring, auditing and persuading, but behind it lay the power to discriminate between oil companies in the allocation of licences. After 1974 the OSO

was attached to the Department of Energy, and in 1975 under the Labour Government, its hand was strengthened by the Memorandum of Understanding which secured the oil companies' formal agreement to the principle of 'full and fair opportunity for British firms'.

The second important policy development was the establishment of the British National Oil Corporation (BNOC) as a state oil company and oil producer. This had important potential for the offshore industry as it was a possible vehicle for selective purchasing. An organisation takes some time to establish. It was not until 1976 that BNOC was set up, and it had only been going effectively for a couple of years when in 1979 the Conservative Government entered office pledged to denationalisation. Britoil was floated in 1982 and is, even aggressively, a commercial enterprise. BP (in the early years over 50 per cent Government owned) saw itself as a British company, but certainly not a vehicle for industrial policy via discretionary use of purchasing power. Further, it was important for BP to emphasise its multinational status in order to secure its position in the United States.

Direct assistance to the industry has been relatively limited. There was a major attempt to use public money in the mid-1970s when the Labour Government, through the Scottish Development Agency, purchased two sites for the construction of concrete platforms. The decision was made at a time when three major contracts for concrete structures had gone to Norway and concrete seemed to offer advantages over steel. The technical judgements were not in fact correct, and one of the sites, although developed at a cost of about £12m, was never used.

The National Enterprise Board[4] played a very small role in the industry but the acquisition of Vickers Oceanonics and the setting up of British Underwater Engineering was undoubtedly an important development. It not only maintained a British capability in an important and rapidly changing activity, but was able in addition to operate as a type of holding company for new high technology firms in the business.

British industrial policy does not discriminate against foreign subsidiaries and indeed, throughout the period, they were welcomed. They were eligible for all the benefits under the Industry Act 1972, under regional policy and any other investment incentives produced by local authorities in the form of infrastructure. Even the OSO made no distinction between British-owned companies and foreign subsidiaries which were operating fully in the United Kingdom. Concern for employment was a major reason for positively welcoming Marathon Le Tourneau into the Upper Clyde Shipyard, GC Doris, in conjunction with John Howards, into the Kinshorne platform yard and Heerema into the Lewis offshore yard in the Shetlands.

The policy of an 'open door' to foreign firms has had major effects on the development of the offshore industry in the United Kingdom. From the point of view of employment and, in the short term, the balance of payments, there is no need to distinguish between British firms and foreign subsidiaries; but in relation to technology transfer and longer-term export prospects, the distinction can be of some importance. In this industry, long chains of contractural relationships are important and it follows that the nationality of the firms which receive the initial or major contracts is often of considerable significance.

The most marked effect of the 'open door' policy was on drilling equipment. The major American firms Combustion Engineering (which is the parent company of both Grays Tools and Vetco), Baker Tools and the National Supply Co. all came to the United Kingdom at an early stage: they acquired remnants of the British drilling equipment industry (which had in fact been quite strong in the immediate post-war period) and today the United Kingdom has no firms of major significance in this sector. But the most important effect of the open door to inward investment was the encouragement of the major foreign oil companies, for these were strongly tied to the American supply industry. Their presence also had an important effect on research policy since it was clear that they could well afford to do their own research and deterred both Government and/or co-operative research effort in the development of major capabilities.

The presence of major overseas subsidiaries may help to explain why British research effort in the early days was so limited. In 1973, the diverse pieces of relevant R&D were brought together by the Department of Industry and in 1975 these responsibilities were transferred to the Offshore Energy Technology Board (OETB) under the Department of Energy. The main burden of their work has been concerned with geological work to support licensing policy, work on safety, and studies of design and materials to assist the development of government regulations of offshore structures and operating practices. A further category is industrial support R&D to help British firms to develop or maintain a position as credible suppliers to the oil industry, but expenditure in this area was only £2–3m per year between 1967 and 1982.

In summary, there was virtually no Government policy towards the offshore supplies industry before 1973. There was no early start and no recognition of future needs. After 1973, when demand was already high, the main policy was the full and fair opportunity arrangements under the OSO. The state oil company was short-lived and the welcoming of foreign subsidiaries made policies which discriminated against 'foreign' firms virtually impossible. The assistance to research was greatest at the basic level and comparatively slight to the (much more significant) development work.

In 1984, in the Eighth round of licensing, R&D contracts placed by the oil companies were included in the OSO monitoring system. This followed the Norwegian example (see p. 233) and reflected growing anxiety about the effectiveness of the R&D policy. In late 1984, the Ninth round conditions included, among the ten factors which would be born in mind when examining applications, the

> extent to which the applicant has involved, or plans to involve, UK owned and controlled organisations in his exploration, development and production activities on the UK Continental Shelf through the generation of new technology, the placement of research and development contracts and the provision of opportunities for the design, demonstration and testing of products and techniques [Gazette Notice, 1985].

This represents a considerable shift in policy in that for the first time there is explicit mention of British-owned companies and greater emphasis on the development of British-based technology via the placing of contracts.

NORWAY

Background
The Norwegian case is interesting because exploration and development on the Norwegian continental shelf has spanned the same period as development in the United Kingdom. The problems have been very similar, but the policy responses are sufficiently different to make comparisons interesting.

The first licensing round in 1965 brought the first major discovery —Ekofisk—in December 1969. It was jointly developed with several nearby fields in the southern part of the Norwegian continental shelf and production from the complex began in 1971. The larger Frigg and Statfjord fields started to produce in 1978 and 1979 respectively. By 1978, oil and gas production (48 million tonnes of oil equivalent) accounted for 17 per cent of Norway's GDP (three or four times the proportion for the United Kingdom) and around one-third of both total export earnings and government revenues.

In the main oil province south of 62 degrees north, the total recoverable reserves of oil and gas are estimated at 4-5,000 million tonnes of oil equivalent. If one includes the necessarily speculative estimates of possible reserves to the north (where expectations have been raised in recent years by successful test drilling), the total resource base of the whole Norwegian continental shelf could be as high as 12,000 million tonnes of oil equivalent. Norway's own oil consumption is only 8 million tonnes per annum. There are only a few well-developed manufacturing industries and industry accounts for only one-third of total employment (two-thirds being in services, agriculture, forestry and fishing). The bulk of Norway's oil production has been (and will be) exported, but Norway has put a limit on production (of oil and gas combined) of 90 million tonnes oil equivalent per annum. This is well above current production levels which are expected to reach 65 million tonnes of oil equivalent in the late 1980s. Even so, the petroleum share of GDP will then be 25 per cent, and the kroner a 'petro-currency'.

Norwegian industry is small and, in general, highly concentrated. Its main strengths lie in shipbuilding, shipping, and in civil and electrical engineering. Several large firms have established capabilities in offshore activities. The major rig construction yards are owned by Aker and Trosvik, shipbuilding groups which seized the opportunity to enter the rig market and were, by the early 1970s, producing their own rig designs. Norwegian shipbuilders such as Olsen, Dyvi and Wilhelmsen showed similar enterprise in making early entry as rig operators; in addition, several other Norwegian shipowners operate rigs. The story was similar with offshore support vessels. Ulstein branched out from making fishing boats in the early 1970s and developed a best-selling design of support vessel. Since the early 1970s a number of ship owners such as Wilhelmsen and Augustsson have built up sizeable fleets of specialist support vessels. In addition to his other numerous business interests (e.g. Timex), Fred Olsen moved into rig-building via his controlling interest in the Aker shipbuilding group, and into rig operation via his shipping companies.

Individual entrepreneurship has, therefore, been a major factor in Norwegian entry into the maritime activities in the early years. The capabilities of Norwegian construction engineers were demonstrated by the successful building of the first concrete platforms in conjunction with G.C. Doris, the French firm.

Industrial policy and performance
The enormous importance of oil to the Norwegian economy, and its impact on the value of the currency, and hence on cost

competitiveness, has led to a highly nationalistic approach. The possibilities of controlling the flow of orders and the advantages to be gained from local experience at the frontiers of a high technology business makes the offshore business an important strategic sector. The concern that oil should benefit the country as a whole is reflected in the determination that the activity should be as far as possible Norwegianised and that particular stress be placed on the high technology areas which are less cost sensitive.

It was recognised that 'Norwegianisation' required the creation of a powerful national oil company to act as the principal vehicle for Government policy and state investment. Statoil, a fully owned state corporation, was created for that purpose in 1972. Through its role as operator on a growing number of blocks, its 'carried interest' participation in all other licensed acreage and field developments, and its responsibility for marketing of the state share of oil and gas production, Statoil has taken a leading role in decisions on petroleum development and has been a major influence in the procurement of offshore goods and services. One of Statoil's important functions has been to give Norwegian firms the experience necessary to compete with foreign suppliers. Statoil has also encouraged joint ventures where necessary to give domestic firms access to foreign technology and know-how and one of its functions is to monitor the extent to which these joint ventures are bringing about a transfer of technology.

Norway was also an early adherent to the 'full and fair opportunity' policy. The second licensing round in 1968 stipulated that there should be state participation in development and production and 'full and fair opportunity' for Norwegian firms. The principle of full and fair opportunity was strengthened in 1972 under a Royal Decree requiring the use of Norwegian goods and services whenever they were competitive.

Norwegian shipowners and shipbuilders were flourishing until the mid-1970s. Unemployment in Norway was negligible and even the 'full and fair opportunity' policy had little operational significance. Norwegian industry did not have the capability to engage in more than a fraction of the early design engineering fabrication and installation contracts but Norwegian firms were obtaining a large share of the contracts for the construction and operation of drilling rigs and support vessels on the basis of competitive tendering. It is estimated that by 1974 Norway was obtaining some 25 per cent of the orders, but this was measured in gross terms and the net value added is uncertain.

With the decline in shipping and shipbuilding which resulted from the world-wide recession from the mid-1970s, the Norwegian Government's policies were strengthened. Concern for unemployment was reflected in large subsidies for the ailing shipbuilding industry. From 1976 onwards the Government required the operators to disclose procurement information to show that Norwegian firms had been given 'full and fair opportunity', and from 1978 the policy was strengthened still further, the operators being required to re-open negotiations with Norwegian construction yards whose bids had been rejected. In some cases it is claimed that the yards then reduced their bids sufficiently to obtain the contracts. The oil companies, for their part, were anxious to obtain licenses and stand well with the Norwegian Government and were therefore concerned to show co-operation with Norwegian firms.

Employment in petroleum-related activities rose steadily from 6,600 in August 1973 to 44,000 in August 1981. At the latter date 2,000 were in the construction yards, 11,000 in offshore work on rigs and platforms, 10,000 at supply bases (including offshore transport and catering and 2,000 in the petrochemical industry). In 1978 the Government stipulated that at least 75 per cent of the workers employed on offshore activities should be Norwegian and, as a result, the proportion of foreign workers fell from the 13 per cent of pre-1977 to 9 per cent in 1981. Foreigners were in effect restricted to jobs such as welding and design engineering, for which there were insufficient Norwegian applicants. The extent to which the policy of Norwegianisation was pushed is indictated by the fact that an American contractor claimed that they were employing Norwegians in parallel with their own people; this undoubtedly tended to increase costs, but it also promoted technology transfer.

The combined effect of increasing Norwegian capability, the increased influence of Statoil and the strengthening of the full and fair opportunity policy led to a major increase in Norwegian participation. For Statfjord A, the domestic share of the final value of the project was around 60 per cent. For Statfjord B, the Norwegian share rose to around 75 per cent. By 1979, the overall Norwegian share was 67 per cent in gross terms (the same basis as the British figures) and 50 per cent allowing for bought-in materials, components and services. By 1982 the gross figure was above 70 per cent, roughly the same as the United Kingdom (but for the United Kingdom the proportion of bought-in materials is lower).

All the early rig and production platform contracts obtained by Norwegian firms were awarded competitively; but the rising Norwegian share after 1976 appears to have been influenced by preferential procurement. Norwegian firms built eight out of fourteen concrete platforms installed in the North Sea (including three in the British sector) and obtained all the concrete jacket contracts and most of the deck contracts through competitive bidding. However, of the twenty-two steel jacket contracts for the Norwegian sector that had been placed by 1979, Norwegian yards received only four and all were awarded preferentially. The number of modules fabricated by Norwegian yards increased steadily, but this, too, was mainly due to preferential treatment.

In concrete construction, maritime services and electric installations, Norwegian costs are thought to have been competitive; but costs in the platform yard industry in the late 1970s were 15 to 30 per cent above comparable import prices and for steel modules, Norwegian costs were 25 to 80 per cent above the cost of importing (MOE Report, 1980). After a searching cost enquiry, an official investigation team placed emphasis on the need to develop Norwegian know-how and skills through participation in the mainstream of international development of new concepts and solutions (MOE Report, 1980). Since the fourth licensing round in 1979, it had been a condition that oil companies applying for new acreage should place substantial research and development contracts with Norwegian firms and research institutes. This policy was strengthened and subsequently a number of oil companies have entered into large research and development contracts in Norway to meet the new requirement.

The new requirement marked a radical departure in petroleum-licensing conditions, both in Norway and world-wide. The R&D projects financed by the oil companies, and chosen by them, are expected to lead to substantial learning by the Norwegian partners and to commercially applicable results. The recovery of the bulk of the 'probable' (but undiscovered) resources south of 62 degrees north will involve both deeper water and more complex reservoirs and will require new technological solutions; north of 62 degrees will probably present even greater difficulties. The Norwegians hope to be at the technological frontier in this type of deep-water work.

By September 1982, the oil companies were committed to paying $230m (1,487 kr.) in the period to 1986 for offshore-related R&D projects in Norway. Nine-tenths of the total was provided by ten oil companies; three-quarters of the total was for work on production, underwater and reservoir management technologies; the other quarter for special projects on the Norwegian continental shelf—especially in secondary recovery from the Ekofisk complex using water injection. Excluding the special projects, the bulk of funding was channelled to twelve Norwegian firms and four research institutes.

The danger of this forced funding policy is that the oil companies will only initiate somewhat peripheral projects and fairly routine research and development tasks. Experience in the United Kingdom suggests that many major development projects involve a massive number of sub-contracts and a wide range of capabilities; it is not clear that Norway can supply all of these. The effectiveness of the R&D requirement is still subject to much debate (Rödseth, 1981).

The international competitiveness of the Norwegian industry is difficult to assess. Exports slumped after the completion of the three concrete platforms for the British sector and in rig building it became difficult to compete with Finland and the Far East. In 1979 income from abroad came chiefly from the hire of drilling rigs (some for accommodation duty) and support vessels. However, more recently, Saga Petroleum has been the operator for a field in offshore West Africa (Benin) and major contracts have been placed in Norway as a result.

Norway has nevertheless had some substantial successes and there is considerable activity abroad by Norwegian firms. Kvaerner has a large engineering firm in London; Aker has obtained hook-up contracts in the British sector, Olsen acquired Lewis Offshore and operated it for several years before selling to Heerema in 1983. Kongsberg, a state-owned armaments group, has successfully branched into underwater control systems and, jointly with Den Norske Veritas, also owns GECO (Geophysical Company of Norway), a leading offshore seismic survey company. This company, and its subsidiary, Geoteam, compete successfully in the British market and on the basis of a virtual monopoly in their home market can do so at low prices. Den Norske Veritas, the certification and registration company, operates world-wide and carries out very large amounts of research.

The biggest differences between Norwegian and British policies lie in the more nationalistic Norwegian policies and their much greater determination to try to ensure the long-term success of what is for them a very important strategic sector. These long-term objectives are being pursued through Statoil (and, more recently, the two other Norwegian oil companies) and the requirement that all the oil companies finance research in Norway and enable Norwegian firms to participate in mainstream development work.

FRANCE

Background
The history of the development of French offshore capabilities starts much earlier than that of the United Kingdom or Norway. In the late 1950s it was labelled a priority sector and since then has received substantial state assistance. It is still regarded as a strategic growth area. The main interest in this section lies in the unique policy style and the instruments used to promote a successful, export-orientated offshore industry.

The French offshore industry is not very large in terms of the number employed; it is estimated that the numbers employed in the offshore related activities rose from about 1,900 in 1969 to around to 10,000 at the height of the French involvement in the Norwegian Frigg field in 1976, and then fell back to about 8,000 in 1979. (Peat, Marwick and Mitchell, 1981). Estimates of this type are extraordinarily difficult because of the amount of sub-contracting and the diverse non-offshore activities of the sub-contractors. However, the small numbers are also due to the fact that France has no true home market and very little involvement in the many labour-intensive service activities. There are one or two 'lead' firms in each of the main offshore activities and some have been highly innovative and have achieved considerable export success (NOROIL, 1982).

In exploration drilling there are two French contractors, Forex Neptune (a subsidiary of Schlumberger) and Foramer; both are comparatively large by European standards (in mid-1982 they owned, in addition to numerous on-shore rigs, eleven and nine offshore drilling rigs respectively). Due to its early on-shore drilling experience, Forex had a good reputation by the late 1950s and was introduced to offshore by Mobil.

In the fabrication of steel jackets, decks and modules, the major firm is UIE (Union Industrielle et d'Entreprise); UIE has been the most successful foreign fabricator in the British and Norwegian sectors of the North Sea. In 1980, UIE also acquired, from Marathon, the American rig builder, an ex-shipyard on the Clyde and it is now building both small jackets and modules in the United Kingdom.

In rig building, the main firm is CFEM (Cie. Française d'Entreprises Metalliques). CFEM collaborated in the development of the 'Pentagone' design of semi-submersible rig in the early 1970s and subsequently became a leading European designer and builder of semi-submersible and jack-up rigs. Like the Norwegians, the French had successful shipbuilding firms long after most British firms had run into difficulties and the adaptability and efficiency of some of the French shipbuilders provided a good basis for their offshore initiatives.

In offshore concrete technology the main firm is GC Doris (Cie. Générale Industrielle pour les Développements Opérationels des Recherches Sous-marines). This firm became prominent around 1970 through its successful development of the Jarlan perforated caisson concept. Doris designed and built the concrete storage tank for Ekofisk, two concrete structures for Frigg and, in association with Howard, a British firm, a concrete structure for the Ninian field in the British sector. Through a licensing arrangement, Doris has subsequently acquired know-how in steel structures. Howard Doris now have a joint venture at the Kinshorne yard in Scotland.

In diving and underwater intervention systems, the principal firm is COMEX (Cie. Maritime des Expertises), which develops, designs, builds and operates a range of manned and umanned systems, ultrasonic testing, video inspection and hyperbaric welding equipment. This firm vies for international market leadership with the American firm Oceaneering. Its success owes much to successful development work done in collaboration with the French navy, to Jacques Cousteau, who undertook fundamental work on deep diving problems and life-support systems and was also a consultant to the international oil industry, and to CNEXO (Centre pour L'Exploration des Océans), a French institute that specialises in maritime problems and technologies.

In offshore installation and pipelaying, the main firm is ETPM (Enterprise pour les Travaux Pétroleum Maritime). In addition, there are two firms, Conflexip and Flexservice, which specialise in the flexible hosing which is increasingly used for field flow-lines and as pipelines for smaller fields. Conflexip developed the concept and produces the equipment, and Flexservice lays and buries it.

These firms form the core of the French offshore industry and are the focus of state assistance in the field. In addition, mention must be made of Schlumberger which is a multinational rather than a French national firm and is the international market leader in well-logging and oilfield services. With an income in 1982 of $6.5bn, Schlumberger has around 50 per cent of the American well-logging market and 90 per cent of the market elsewhere; over the past few years it has bought Fairchild and several other electronics companies (electronics is of major importance to well-logging technology but this purchase should be seen as part of a major diversification into the whole electronics field). It is an interesting example of a firm which developed expertise and became a multinational company with numerous foreign subsidiaries (its French operation is now essentially one of those subsidiaries). There are analogies here with BP in Britain, especially after it had expanded its operation in the United States. Such companies illustrate the point that when a company has grown into a multinational company, especially one which is diversified in terms of products, markets, and location of manufacturing, it is difficult for a Government to use it as a vehicle for promoting nationalistic policies.[5]

More recently, parts of the French industry have been in considerable difficulty; AMREP, a holding company whose main subsidiary is UIE (see above) has filed for bankruptcy and its main activities have been taken over by Bouygues, a leading French private construction company; the purchase also includes a stake in G. C. Doris (see above) (*The Financial Times*, 15 January 1985). Further, Technip, the main French engineering group owned largely by the Government and the state-controlled oil companies, made consolidated losses of 224m French francs ($29.4m) and expects its operating losses for 1984 to be around 250m frs ($27.9m). The consolidated deficit will have been further swollen last year by losses incurred by Creusot-Loire Enterprises, its civil engineering subsidiary, which is now liquidated. Further finance is being put into Technip by the Government and other subsidiaries. The policy is, as in the past, to ensure the future of such important companies. Technip obtains some 80 per cent of its earning from exports.

The policy framework

Before examining the specific forms of state involvement it is necessary to point to two general and persistent features of French industrial policy. The first is the Colbertian tradition of state support for key industries which since the era of de Gaulle has focused upon selective assistance to lead firms. Lead firms are seen as being of great importance because they generate demand for specialised components and materials. The French emphasise the importance of *'filière'* — the connections between firms which create opportunities and markets both upstream and downstream. The French industry has also derived strengths from the administrative style which makes it possible for the state and industry to work together effectively; the system is conducive to technocratic and technological decision-making.

It has been a long-standing French policy to become independent of the international oil companies and to generate capabilities in the oil industry. French oil companies were first favoured in 1948.

Shortly after Liberation, the Government recognised the need to promote the development of petroleum technologies and in 1945 the Government created the Bureau de Recherche Petroleum and the Institut Français du Pétrole (IFP) (Saumon and Puiseux, 1977). The discovery of Algerian oil promised a major secure source under direct French control and in the 1950s the Government pushed the national oil companies and equipment suppliers into taking the main role in developing Algerian oil—so much so that the major American oil companies found it expedient to secure French partners for their Algerian activities. Unlike the British and Norwegian industries, the French industry had early onshore drilling and development experience with the large Lacq gasfield in France as well as in Algeria.

Algerian independence in 1962 meant the loss of direct French control over a large source of supply and although Algeria agreed to maintain the level of supplies for fifteen years, anxieties about the security of supplies and of dependence on the American majors remained and there were significant policy initiatives. The two nationalised oil companies, Total (CFP) and Elf-Aquitaine (SNEA) were required to develop sufficient overseas sources of supply to expand in line with French demand; the finances available to the Institut Français du Pétrole (IFP) were increased and the Comité d'Étude Pétroleum Marine (CEPM) was set up. These policies both required that Total and Elf should keep abreast of advances in all aspects of petroleum technology, including that required for offshore exploration and production, and provided the resources and the organisation to assist them to do so. Very close relationships grew up between the oil companies and the 'lead' firms of the French offshore supplies industry. In 1965 the national oil companies interested a number of firms in participating in development efforts, anticipating moving into offshore work off West Africa, the North Sea and to a lesser extent elsewhere.

Total and Elf have in effect carried the flag for the French offshore supplies industry. Around 40 per cent of the value of contracts obtained by French industry during the 1970s arose from projects where the French oil companies were operators (Peat, Marwick and Mitchell, 1981). A high proportion of offshore exploration and development contracts in West Africa, especially Gabon and Cameroon, were obtained by French firms through the French oil company operators. The same was true with the Frigg field, which gave the French oil companies and offshore suppliers experience in developing a major, technically demanding North Sea project. The extent of the French participation was of course limited by the British and Norwegian policies which gave their own industries considerable preference.

French policy depends very largely on the close relationships between state, national oil companies and the leading firms. The Institut Français du Pétrole is central; by the late 1950s, IFP was promoting French drilling expertise and organising training courses. Subsequently, it became one of the world's leading petroleum research institutes, with an income in 1981 of 738m frs and a staff of over 1,600 (plus 200 trainees). Over 70 per cent of the income is financed from a special tax on oil and the remainder comes from external contract research and fees from the extensive licensing of process innovations in France and abroad. IFP's governing board includes representatives of government, Total and Elf, industry and the government research funding agency, Centre National de la Recherche Scientifique. In addition, Total and Elf sit on IFP's research steering committees to ensure that the research is relevant to their operational experience and requirements.

IFP also controls a holding company, ISIS (Internationale de Service Industrielle et Scientifique). ISIS has formed a dozen subsidiaries that specialise in different aspects of petroleum development, engineering and equipment supply. Each subsidiary was created to commercialise on IFP's research and retains access to IFP's ongoing research. Formed in 1956, the ISIS group, by 1981, had 6,000 employees and a turnover of 3.5bn frs ($645m). Its subsidiaries comprise three engineering consultancy companies, four design and construction companies (including, in the offshore area, G. C. Doris), and five equipment and service companies (including Conflexip and Flexservice).

The means of co-ordinating joint development programmes and channelling government money into research and development is the Comité d'Etude Pétroleum Marine, which has representatives of government, national oil companies, the offshore contractors and IFP. CEPM was created in 1963—a year after Algerian independence—to support the technology development needed for the French oil companies' move into offshore exploration and development, initially in West Africa (Delacour, 1981). Joint programmes involving both sides of the French offshore industry and the major research institutes were established (in 1963) to develop new offshore exploration and production systems. By 1968 the work had progressed sufficiently to yield a range of new designs: Pentagone semi-submersible rig, the 'Pelican' design of drillship for deeper water and a floating production system incorporating an articulated column, and extensive trials of manned and unmanned submersibles had been completed. In the ensuing years, commercial orders were placed for Pentagone and Pelican designs, COMEX rose to prominence as an international diving contractor and the Pelican was superseded by a more advanced drillship design, the 'Pelerin'.

Through CEPM, the Government spent 400m frs on offshore technology development from 1965 to 1974; this represented 54 per cent of the total national effort and most of the remainder was paid by the national oil companies. By the time the discoveries in the northern North Sea began to be developed, the French were prepared for many of the technical problems and the Ekofisk and, more particularly, the Frigg projects provided important opportunities to demonstrate the effectivenss of French know-how.[6]

In 1975 CEPM began its 'deep sea' programme which concentrated on deeper water exploration and production concepts, pipeline and pipe repair, fatigue in steel and concrete structures, and trials (in offshore Gabon) of a floating production system and submersibles at depth beyond the reach of conventional diving methods. A further phase (programme hydrocarbure française) began in 1980 with the principal aim of commercialising the techniques developed earlier and establishing capabilities to drill in waters some 10,000 ft. deep with a view to exploring the French sector of the Mediterranean.

To sum up, French policy has three major features. The first is the foresight and determination with which the French embarked on a policy of developing offshore technology, including technology suitable for the North Sea, in the early 1960s. The second feature is the very close relationship between the Government and the national oil companies. This was particularly important because it enabled the French to pursue a purchasing policy which guided the development of the industry. Third was the extent to which the development of technology derived from co-operation between research, the client and the manufacturer. The policies have met with very considerable success, both in terms of exports and foreign operations, but there is nevertheless a degree of nationalistic isolation and continued success depends crucially upon continuing development and maintaining industrial efficiency.

HOLLAND*

Background

In Holland, the offshore industry had closer antecedents than in the other European countries. The Dutch civil engineering industry,

* This section owes much to Cor de Feyter who was writing a paper for the Industrial Council of Oceanology (IRO) during the period he spent with SERC (September to November 1983).

and particularly the 'wet' civil engineering industry (which undertakes the construction of dykes and harbours), has a long history and was strengthened by the massive amount of work which followed the great flood of 1954. There were major works in the enclosure of Zeeland and the filling of Ijsselmeer and there was also the upgrading of the Rotterdam Port and the re-organisation of the Scheldt Delta. Further, with the growth of Rotterdam, the process engineering industry had a great deal of experience in petrochemical work. Firms in both these activities had experience of working abroad, particularly in the Middle East and Far East. The Dutch shipbuilding and shipping industries were also strong, particularly in barges and salvage work. The first jack-up rig to be built in Europe was built in 1957 in Holland by one of the five shipyards which were later merged to form IHC. Oil and gas production began on a small scale soon after the War. NAM, the Oil Corporation of the Netherlands, was set up in 1947/8 as a joint venture of Shell and Esso, and is very closely connected with the Government. The great Gröningen reserves were found in 1959 and they still constitute over 95 per cent of Dutch gas reserves. Offshore gas production started in 1970 but oil (in small quantities) was not piped from offshore until 1982. Much of the Dutch offshore industry today works in the international market rather than the 'home' market.

The 1960s were for Holland a period of confident and rapid growth and industry was well poised to seize the oil and gas opportunities in the North Sea. That they seized these opportunities at an early stage is well illustrated by the number of firms that got involved in the industry, even as early as 1962/3. The Wilton-Rotterdam Dockyard Group (WF–RDM Groupsmaatschappij B.V.) became part of the Rhône Schelds Verole Corporation (RSV), the shipbuilding conglomerate which was building modules and small jackets on a considerable scale by 1966. The de Gusto shipbuilders went in for design work and were employed as designers and builders for the first French drillships in the mid-1960s.[7] In 1966, the Dutch Government sponsored the establishment of Sedneth, a joint venture drilling firm between two of the leading construction companies—Royal Boskalis-Westminster and Hollandse Beton Group (HBG)—and Sedco, one of the leading American firms. And it was also in the late 1960s that Pieter Heerema came back to Holland (from Venezuela) and embarked on a joint venture with Brown and Root. He quickly broke with them and after two years was ready to start operations with a new heavy-lift vessel. This firm has grown to a very prestigious position in the industry, its success being attributed to the technical and commerical excellence of the founder.

The large majority of companies now active either in the North Sea or in other offshore areas (world-wide) existed before 1970 (the first year in which gas was delivered to the Dutch coast from the North Sea). Of the 230 Dutch firms entered in the IRO Catalogue as operating in the North Sea, a mere fifty were established after 1970.[8] Some of the firms active in this area are very large and diversified. Royal Boskalis-Westminster, Volker-Stevin, IHC, Hollands Beton Groep (HBG), Ballast Nedem, and Nedlloyd are all very substantial companies.

Besides possessing a suitable industrial infrastructure, the Dutch also had a major research and development input. TNO is an enormous research organisation: by the early 1980s its annual cost was of the order of £300m a year and, together with a number of other institutions, it provides the basis of the Dutch effort to maintain a strong position in high-technology activities. It is extremely difficult to assess the importance of TNO; it has been criticised in recent years for becoming too remote and academic, but this is a common criticism and few firms seem to appreciate the contribution it makes to basic research. Indeed, it appears to be regarded as a free good. TNO also makes a substantial contribution to the training of personnel and to the increased confidence with which people tackle issues at the high-technology level.

Offshore demand in Holland was superimposed on other demands for similar products. Offshore is not a sector but a market. The companies which have specialised in equipment for offshore operations had their roots in shipbuilding or structural engineering. Often they have also continued carrying out their former activities as well. 'It is impossible for a company to say we have had a hundred men working in shipbuilding and 70 for the offshore. That can change from one day to the next'.[9] Total North Sea turnover for the two years 1981 and 1982 is estimated at some $10bn per year. The Dutch offshore industry estimates that their turnover averaged $1bn over the last five years and although it was not exclusively concerned with the North Sea, it probably obtained just under 10 per cent of that business. However, these figures are gross figures and it is estimated that the value added reached only about 30 per cent of the contracts obtained.[10]

THE DEVELOPMENT OF THE INDUSTRY AND
GOVERNMENT POLICIES

The most important characteristic of the Dutch economy is its very high dependence on international trade, which makes up some

50 per cent of GDP (compared with some 25–30 per cent in the United Kingdom). Of that trade, 30 per cent is with Germany and 20 per cent with France and Belgium. Largely as a result of its dependence on trade (but in part a cause of that dependence), the Dutch have a very open and competitive economy.

As far as the offshore industry was concerned, the Dutch were, in the period up to 1975/76, remarkably successful. Having made a good start in the 1960s, Royal Boskalis-Westminster, Nedlloyd, Schmidt International Fugro (sub-sea surveying), Heerema, deGroot and HBG all became important in the industry. HBG's entry into module fabrication was relatively late; they started during the 1973/74 boom, having been encouraged into the business by Shell. Compared with the major shipyards, it was a relatively stripped-down, flexible organisation. The other major module builder, Grootint, expanded from work in the process plant industry; it grew rapidly and its overheads were also relatively low. As module fabrication became increasingly competitive, price competition was important and it was firms with low overheads that competed most successfully for the contracts.

The 1970s, however, brought economic problems for Holland. Gas revenues boosted foreign exchange earnings but also raised the value of the florin, which in turn hit the competitiveness of Dutch industry. Substantial sums went in selective help for firms in trouble, but there was also much discussion of the need to give less Government assistance to large, weak and possibly incompetent firms. There was particular concern about the amount of money being put into the shipbuilding industry and into RSV in particular.

In 1980 anxieties about the economy were reflected in a 'gentleman's agreement' with Shell and Esso (joint owners of NAM). In a rather complicated arrangement concerning gas prices and their relation to oil prices, it was agreed that NAM's profits, which were increased by the new price arrangements, should, in part, be re-invested in the Netherlands. The undertaking was that $2bn per annum should be set aside for projects which would help to restructure Dutch industry by improving its technology profile. These investments—though mainly energy-related—were to go to industries both within and outside the oil or gas industry.

The arrangements were surprisingly detailed. Both Esso and Shell would invest $1.4bn each in the years to 1984 in exploration, recovery and transportation of natural gas. Moreover, Shell would invest in refinery and chemicals for a total of $2.1bn and Esso would invest $1.15bn in coal and coal-related technology. Esso would also add $1.4bn to refinery investments in Rotterdam. Investments, considered for the period from 1984 onwards to 1989, total $11bn. Following postponements of investments in the coal gasification sector, both companies proposed to invest in a more fragmented way, i.e. in a large number of smaller projects. Both companies increased the number of firms on their 'qualified vendors' lists; this, particularly in the case of Esso, substantially improved its relations with the Dutch industry.[11]

Part of the agreement was a regular series of meetings between top rank civil servants and officials from Shell and Esso. These meetings were planned to enable Government officials to use information thus acquired to make sure that Dutch firms participated in the programmes under consideration. Given the preparation at ministerial level, 'the implementation phase of the gentleman's agreement thus was proceeded in an environment of compatible authority'.[12]

In June 1980, the Netherlands' Scientific Council for Government Policy published a gloomy report entitled 'The place and future of industry in the Netherlands'. The report was very largely concerned with the problem of the choice between general policies and sectoral policies. Unemployment was increasing in Holland (it reached 14 per cent by 1982/83). In 1981, the Wagner Commission was set up. The chairman was the chairman of Shell, and both Philips and Unilever representatives were on the committee. This committee appears to have been searching for an escape from the policy of giving large help to weak companies. The main recommendation was the setting-up of the Company for Industrial Projects; this was done promptly and the new organisation was in place by September 1982. A small number of projects, including one with Heerema for a new jacket concept, had been agreed.

This policy bears some relationship with the French approach but is in sharp contrast to the main United Kingdom policies. It may be described as a 'top down' policy: the idea is that if large projects involving advanced technology are undertaken they will give rise to demands for further research and to sophisticated orders for sub-contractors and, further, that the advanced designs will lead to export capabilities. It is thus an application of the French concept of industrial *'filière'*. 'Bottom up' policies, which involve improvement of the quality of basic research and sub-contractors are complementary rather than competitive with 'top down' policies; the enormous provision for research is the major part of the 'bottom up' policies already in place. It is, however, significant that the Government is making every effort to bring together the research organisations and industry. Not only are they asking the research organisations to obtain a much higher proportion of their own finance, but they are also concerned to promote co-operation so that the research organisations will have a better indication of the relevance of different lines of research. In other words, there is also increased emphasis on 'top down' policies.

Dutch policy is based, essentially, on competition.

> In our country, an open door policy is practised. We encourage the oil and gas industry to make use of local products and services on a competitive basis, but without restriction on outside companies. And, without restrictions includes that it is not necessary to trade through a Dutch-based venture.[13]

Nevertheless, since 1980, policies have concentrated on close co-operation between the Dutch Government, Shell, Esso and other leading companies. Emphasis is placed on working with leading clients rather than the firms which it is hoped will receive the sub-contracts.

The openness of the Dutch economy and the international involvement of many of the leading Dutch firms has resulted in a number of the firms, notably Heerema and Royal Boskalis-Westminster, becoming true multinationals. They are therefore less available to the Government as instruments for industrial policy. Shell and Esso's ownership of NAM and the Government's control over oil and gas policy makes these two oil companies very important. Although there has been a slight shift towards top down policies the scope is limited because, in an open economy, particularly one with such major trading relations with West Germany, a high domestic content cannot be guaranteed and the quality of the industrial base is therefore crucial.

Assessment of the policies

THE POLICIES PURSUED

Our examination of the main policies pursued shows considerable differences between the approaches adopted in the four countries. Requirements of the offshore industry were common to all, but the industrial background, the newness of the activities and the economic/political backgrounds varied considerably. In varying degrees all governments were concerned to encourage the growth of the new activity. We have identified six types of policy and these are set out in Table 7.3, which also summarises the use made by each country of the different policies.

Below we discuss the six types of policy and the circumstances

Table 7.3 Summary of policies used in Britain, France, Norway and Holland

Type of policy	Britain	France	Norway	Holland
1. Assisting early entry	Negligible	Substantial	Slight (successful private firms)	Negligible (successful private firms)
2. Long-term major project R&D	Negligible	Substantial	Increasing in recent years	Increased recently (substantial basic research)
3. National oil company purchasing	Slight	Major	Major	Slight—through NAM (Shell/Esso)
4. Full and fair opportunity	Major	Policy not available	Major	Policy not available
5. Open door to foreign firms	Major	Strong restrictions	Controlled joint ventures	Open door
6. Selective promotion of major firms	Negligible (except for platform construction yards)	Major	Significant	Minor—but increasing

Source: Government Policy for the Offshore Supplies Industry, Britain compared with Norway and France (1983), Lesley Cook and John Surrey, SPRU Occasional Papers Series No. 21, Brighton, University of Sussex.

in which the different countries relied upon different strategy combinations.

(i) *The timing of Government involvement*
France began to pursue long-term, forward-looking strategies in a number of the new high-technology industries, notably nuclear power, soon after the War and was very actively engaged in offshore work by the early 1960s.

(ii) *R&D support for major projects*
A distinction should be made between support for research institutions and support for major development work. The French supported project development on a large scale and the Italians also did major development work within their nationalised industries. Early large-scale development work gave rise to operating experience and know-how. It is, however, quite unclear whether, fifteen years on, the advantages to be obtained from large-scale development work are comparable. The industry is now much more complex and it is doubtful whether large-scale projects would now give rise to such decisive advantages.

(iii) *Using oil companies' purchasing power*
France, Norway and Italy have pursued this policy. With close co-operation between the government and a state oil company, the company becomes in effect an arm of government. It becomes an informed and experienced buyer and develops real knowledge of the suppliers: it is able to place contracts in order to develop competence, to give a chance to firms to prove themselves and to co-operate in development work which it knows to be of importance. An efficient oil company can (French style) help firms to become more competent and, by developing a collaborative long-term relationship, can help to engender a long-term strategic view of development. The importance of basic efficiency cannot be overstressed. It appears that the French have, within a highly protected system, been efficient, but the Norwegians have had major doubts about their performance (MOE, 1980). The advantage of an oil company is that, in a rapidly-moving innovative situation, it can make judgements which it is difficult for any department of government to make. In the United Kingdom, Norway and Holland major efforts are being made to bring about this type of co-operation between the clients and industry but this can be easier to organise when those clients, the oil companies, are an arm of government.

(iv) *Full and fair opportunity*
The policy of full and fair opportunity is only open to those Governments which have some means of inducing the oil companies to comply. Both the United Kingdom and Norway have made this an important aspect of policy. It is more effective when what is needed is to ensure that indigenous firms enter the new activity, and less effective when it is a matter of entering, not a relatively stable business, but one in which there is continuous innovative change and a need for undertake longer-term development.

(v) *Foreign subsidiaries, take-overs and joint ventures*
Government attitudes and policies towards foreign ownership and the subsidiaries of foreign-owned companies differ markedly between Britain, France, Norway and Holland. France pursues the most nationalistic policy. Although a number of American equipment manufacturers have French subsidiaries, there is a strong desire to promote French firms with advanced technological capabilities. As in other areas of French industrial policy, take-overs are curtailed, assistance to subsidiaries is highly selective and joint ventures operating in France are comparatively rare. The Norwegians have relied heavily on joint ventures but wholly-owned subsidiaries of foreign firms are discouraged; the general policy is one of Norwegianisation combined with technology transfer through joint ventures. These policies are in sharp contrast to the British open door policy which encourages foreign inward investment, permits foreign take-overs in major industries and precludes discrimination between foreign and British firms in the granting of investment allowances. In the administration of the full and fair opportunity policy, via the OSO, foreign subsidiaries employing substantial numbers are regarded as British.

The effect of the United Kingdom's open door policy is difficult to gauge. This is an industry which is essentially an extensive network of sub-contractors. If a foreign subsidiary is established, the impact on British industry and employment depends partly upon its purchasing behaviour. At one extreme, foreign subsidiaries might purchase from their home suppliers and do all the technical development work in their home country; at the other extreme, they purchase largely in the foreign country where they are operating. Much depends on the efficiency of the local sub-contractors—the major international contracting firms must purchase from the best source to maintain competitiveness. Overall, it would appear that the open door policy has had three important effects. First, it has stimulated many sub-contractors and enabled them to learn the requirements and technologies needed in the international oil industry. Second, the strength of the experienced multinational competition has inhibited the development of indigenous firms which might have become leading contractors or equipment suppliers. Third, the existence of many foreign firms in key areas in the offshore industry has made it difficult and unattractive for the British Government to promote long-term development work in Britain; large parts of the benefits of such assistance would have accrued to the parent companies and been exploited world-wide.

(vi) *Selective promotion of lead firms*
The promotion of lead firms has been a central aspect of French industrial policy and is closely associated with their concept of the *'filière'* as the focus for long-term technology policy. The Dutch are making a major effort to link their technology development work with the needs of leading firms and in this sense are now promoting leading firms. In Norway selective treatment of the larger firms has been the consequence of the small size of the economy. There are few large firms and it is administratively and politically simpler to deal with them individually. In Britain the selective promotion of leading firms has hardly featured in policy. This is partly because there were numerous potential entrants and partly because those that did attempt entry found the going very tough (they had after all to compete from the start with the international giants). The case closest to a 'lead firm policy' was the assistance of Vickers in sub-sea engineering and small manned submarines; great hopes were placed on this until, with the increased use of remotely operated vehicles (ROV) and intense competition, Vickers withdrew from this area of business.

POLICY IMPACT IN VARIOUS SECTORS

The heterogeneity of the offshore industry means that the impact of policy really needs to be assessed, not for 'the industry' as a whole but sector by sector. Assessment of policy impact is difficult because it is important to avoid simple *post hoc propter hoc* deductions. The success in different sectors of firms in different countries depends as much upon the history and the efficiency of the firms as on Government policies. First we consider the probable impact of the policies pursued on (i) output and employment and (ii) national firms. This assessment is based on policy analysis and not upon the actual outcome. We go on to discuss the extent to which the development of the different sectors in the four countries is in accordance with the probable effects of the policies.

(i) Output and employment

Table 7.4 summarises our judgements, based on the study of their main characteristics, of the probable impact of the different policies on output and employment in each of the main sectors of the industry. Purchasing by state oil companies is particularly important for all the main offshore specific activities, where cumulative experience is needed. Full and fair opportunity is a strong policy for major sub-contractors in fabrication and equipment, where scale, accumulated experience and R&D lead times are less significant. In contrast, early entry and major R&D are important for promoting firms in the major services of installation, pipelaying and exploration drilling and for firms making sophisticated drilling equipment.

(ii) National firms and foreign firms

Table 7.5 presents judgements on the different effects of policies on national firms as opposed to foreign subsidiaries. The first important point is that the combined effect of full and fair opportunity and the encouragement of foreign subsidiaries tends, when the activity is new to the country but not to the world, to bring in experienced multinational companies. Second, when long-term R&D, scale and experience are important, it is increasingly difficult as the work becomes more sophisticated to compete with firms which have entered early, gained experience, and established a strong position. Third, in an industry in which much of the work is sub-contracted, the oil companies and major lead firms responsible for design and the placing of contracts are important for the development of the many sub-contractors. Thus, when foreign lead firms are superior to indigenous firms their presence is likely to benefit the many sub-contractors.

THE PATTERNS OF DEVELOPMENT

A number of interesting points emerge when Table 7.5 is considered alongside Tables 7.3 and 7.4. Having relied upon 'full and fair opportunity' and pursued an 'open door' policy to foreign firms, Britain presents a picture much in line with that to be expected if policies alone determine the outcome. No leading firms have developed in sectors where large-scale R&D and learning are important; a high proportion of the British sector fabrication work has been obtained; and considerable strength has been built up by firms producing equipment adapted for offshore oil production. The main divergence is the strength, without substantial assistance, of the design and

Table 7.4 Effects of Government policies on output and employment in the different sectors

Main sectors	Early entry	Major R&D	National oil company purchasing	Full and fair opportunity	Encouragement of foreign subsidiaries	Chosen firms' preference
Design and management engineering	*	*	+	*	*	−
Fabrication	−	−	+	+	*	−
Equipment						
− Oil specific	*	−	*	+	*	−
− Adapted for offshore	*	−	*	+	−	−
Major services (pipelaying and installation)	+	+	+	*	*	*
Drilling and exploration	+	+	+	*	*	*
Underwater engineering	*	*	*	*	*	*

* Very important in terms of work and employment obtained in the sector.
† Fairly important in terms of work and employment obtained in the sector.
− Generally unimportant.

Table 7.5 Sectoral impacts of the policies used in the three countries

	Britain	Norway	France
Policies used			
Early entry			
Major R&D			
National oil company purchasing			
Full and fair opportunity			
Encouragement of foreign subsidiaries		(Joint ventures only)	
Chosen firm preference			
Sectors assisted to obtain work	All, but especially fabrication and equipment of both types	All, but especially civil engineering and fabrication	All, but especially exploration, rig-building, underwater engineering, pipelaying and installation
Sectors in which national firms have benefited most	Fabrication and adapted equipment	All, but especially civil engineering and fabrication	All, but especially exploration, rig-building, underwater engineering, pipelaying and installation
Sectors in which foreign firms benefit most	Design and management engineering; oil-specific equipment; pipelaying installation	None directly, but long-term results of joint ventures are unclear	None

management engineering sector. In Norway the development is somewhat less closely related to the expected policy-related pattern. The big difference is the considerable early success in rigbuilding and drilling. This success, particularly in rigbuilding, must be largely attributed to entrepreneurial initiatives and relatively cheap finance.

In France, the emergence of a number of firms with strong offshore capabilities in the major activities, as in Britian, have been in line with the pattern which could be expected to result from the policies pursued; however, the degree of success—particularly in high-quality fabrication, underwater engineering and exploration drilling—owes much to the efficiency of the firms involved.

Conclusions

The development of the North Sea gas and oilfields made new demands on the oil companies and on existing firms in the industry— the firms engaged in offshore work in other parts of the world and those primarily engaged in onshore work. It was a question of developing technology and expertise to meet the ever tougher conditions of the North Sea as exploration developed and pushed into deeper and more northerly areas. European firms, coming from other industries but new to offshore, had much of the general expertise required but there was also much to learn; it was a matter of learning to do what was already being done by American firms *and* continuously developing their capabilities to keep abreast of advances.

The distinction between learning to do what is already being done elsewhere and acquiring the capacity to develop further is of great importance. The 'infant' industry metaphor applies mainly to the former because it implies that maturity is a recognisable and fairly static state. The capacity to develop continuously is much more demanding than the ability to do what is already being done although, in most cases, the latter is an essential prerequisite. Only when there are major technological discontinuities is there the option to leap-frog existing technology. This is not the case for any major sector of the offshore industry; technological change is largely embodied in complex systems and is incremental. The new demands of the North Sea required both knowledge about how things were done and a capacity to solve problems and overcome difficulties.

The success of European firms in this industry depended partly upon making the initial adjustments to offshore work and partly on their ability to cope with this dynamic adjustment process. Some sub-contractors, such as the fabricators and those supplying adapted equipment, had to adjust to new requirements and standards, and many firms had to solve new problems. The smaller sub-contractors were presented with fairly specific problems, but the oil companies and the major contractors needed the ability to cope with the dynamic adjustment process over a wide range of technologies. Two issues crop up again and again in the discussion of the development of the industry: the first is the role and importance of 'lead firms' and the second, the direction and value of research and the importance of technical development.

THE CONCEPT OF LEAD FIRMS

The concept of 'lead firms' has two interpretations: a 'lead firm' may mean the oil company operators and the major contractors who 'lead' in the sense that they decide what to do and where to place the contracts. This interpretation is useful in the offshore supplies industry because the size and complexity of contracts gives rise to a great network of sub-contractors. The second interpretation is that of the technological leader in a situation of progressive change. Many firms are 'lead firms' in both respects. These include the major oil companies, both because they determine what shall be ordered and from whom, and because they develop many of the new concepts. Some of the major installation contractors, notably Brown and Root and McDermott and, more recently, Heerema, would be classed as 'lead firms' in this sense. These are large powerful multinationals and their existence and their influence on the many sub-contractors has been a major determinant of policy. For the main types of equipment, such as downhole drilling and well-head equipment, the 'lead firms' have a technological lead, and often a considerable degree of monopoly. Again, many, notably Combustion Engineering, Halliburton and Schlumberger, are multinations with great technological resources.

POLICIES CONCERNED WITH 'LEAD FIRMS'

The policy issues are whether some national firms can be picked as potential winners or, if necessary, rescued and given a second chance. In France the close connections between the Government, the state oil firms and the industry made it possible to give preference through nationalistic purchasing policies and R&D support and effectively to nurse potential lead firms to a position of strength. In Norway, where the choice between indigenous firms was limited, reliance was placed on joint ventures and protection, initially to establish lead firm capabilities and subsequently to develop these capabilities.

In the United Kingdom, while there was no explicit decision that it was impossible (at reasonable cost) to create lead firms from indigenous suppliers, the policy pursued effectively sabotaged such a strategy. By encouraging major multinationals to establish themselves in the United Kingdom and treating these firms on a par with domestic industry, it was made very difficult for indigenous firms to develop.

The reasons for this are worth spelling out since they throw light on the role of the lead firm and the sheer difficulty of 'creating' them.

First, direct competition with large established firms is difficult when, as in the offshore industry, technology transfer is a matter of transferring a wide range of experience and expertise and not, as is sometimes suggested, a matter of obtaining a discrete piece of information. The multinationals sent over considerable numbers of people to set up their new subsidiaries and, particularly where design and new project work is involved, also relied upon advice and help from the parent company. Experienced firms had much of what was needed for the new North Sea work and were able to go forward quickly. Without either the early start of the French or the preferential treatment of the Norwegians, competition with the multinationals was bound to be difficult.

Second, although entry into the offshore construction business was relatively easy in the early 1970s when demand was very high, growth and the development of major capabilities was much more difficult. Lead firms undertaking large projects incurred great risks. Many major firms both in Europe, and indeed in the United States, have incurred large losses. In the United Kingdom there were many firms competing and, although this eliminated the inefficient, it also tended to weaken firms which otherwise might have succeeded in developing successfully. The more potential major firms there are and the lower their efficiency as compared with foreign firms, the more difficult it is to pick winners or to identify 'lame' (as opposed to dying) ducks. Moreover, the absence of a state oil company which could (as in France) help nurse potential candidates, compounded these difficulties. It was left to the market and no indigenous lead firms emerged; instead, many major firms since 1973 have withdrawn, nursing heavy losses.

The success of the (relatively few) leading European firms suggests that they must either develop on the basis of a major technological success (for example, Heerema and Schlumberger) or have some degree of preferential treatment, as in the cases of Forex, ETPM, UIE, COMEX, Kvaerner, Aker, Samprojetti and Saipen.

R&D POLICIES

Assistance for research and development has been a significant part of all European Governments' policies for the offshore supplies industry. The objective, although partly the general increase in knowledge to secure adequate safety regulations, has increasingly been to assist firms in the supply industry to compete and to obtain commercial orders.

Research and development in the offshore supplies industry is a complex activity with many 'players' and many 'levels', ranging from fundamental research to major semi-commercial prototypes. The importance of both an efficient research infrastructure, supplying ideas and information, and efficient systems for creating a demand for useful research, is seen very clearly in the offshore supplies industry. Concept research often stems from the major firms themselves or the research institutes with which they are closely associated, but at the same time there are many laboratories and firms touting new ideas to potential users. Development, which is so important in this industry, is mostly done by a considerable network of sub-contractors rather than by a single firm or laboratory, much of it initiated by firms buying in research by placing R&D contracts. Thus, in this industry, R&D is being *both* pushed from the supply side (people with ideas) *and* led from the demand side (problems to be solved).

Government involvement raises two issues. The first is the question of what research and development Governments should assist, given that the results will very soon be available to all. This 'free rider' question was peculiarly acute for the United Kingdom, which had welcomed the multinational oil companies and the foreign subsidiaries, the result being that publicly-funded research is of prime benefit, not to British firms, but to wealthy multinationals.

The general response to this problem has been to concentrate on basic research and applied basic research, increasing the funding of research institutes, universities and the government safety authorities for work relevant to the offshore industry. France, which has in effect spurned the multinationals, has been less influenced by the 'free rider' problem. From an early stage, publicly-funded R&D has played a central role in RFrench strategy—the funding covering both basic research and major project development work.

Major commitments were made to this programme from 1963 onwards. In the circumstances, it is not surprising that in the two most open economies, the United Kingdom and Holland, R&D policies are concentrated mainly on the supply side and in particular on the basic research infrastructure. France, although providing substantial resources for basic research, placed much more emphasis on the demand side.

The second issue arises over this 'demand/supply side' division. On the part of Governments there is increased recognition that concentration on the supply side of research and development infrastructure can make it difficult to use the resources efficiently for the promotion of commercial capabilities. The problem can be reduced by inducing the research laboratories to seek a proportion of their funding from commercial contracts—i.e. to sell their capabilities more effectively. The promotion of demand-led research is, however, somewhat more complex institutionally; as we have seen, attempts to increase 'top down' initiatives were introduced in Norway in 1979, in Holland in 1983 and in the United Kingdom in 1984. All involve collaboration with the oil companies and other lead firms. Without the French capacity to influence a number of key firms in the *filière*, it is proving more difficult than originally thought. But the measures have only been tried for a very few years, and it is uncertain either how best to pursue this policy and/or how much progress along this route is possible. It is certain, however, that there is much to learn and that this is an important route.

THE SUCCESSFUL PROMOTION OF NEW
INDUSTRIAL ACTIVITIES

What lessons, if any, can be drawn from this brief history of the offshore industry in relation to the promotion of new activities? The four country studies show very different patterns of development and very different Government policies. The United Kingdom has an industry which is large, but foreign-dominated, in that most of the contracting and the technological lead firms are foreign. Many are, however, fully established in the United Kingdom and operate very much as the equivalent indigenous firm would. A large number of British firms have become good 'second rank' firms and many have been successful in particular niches. There is little doubt that the establishment of the major multinationals (including the oil companies) in the United Kingdom and the experience gained by indigenous firms working with them has increased the efficiency and the export potential of many of the sub-contractors. On the other hand, the unhappy experiences of many of the large British firms which entered the industry, some of which might have become lead firms under a protectionist policy, makes it quite uncertain whether, on balance, a policy of promoting lead firms through early and substantial R&D support and preferential purchasing would have been successful. In an industry with extensive sub-contracting, protected but inefficient lead firms result in the worst of all possible worlds.

The French, with their early start, developed a number of indigenous lead firms, Policy was designed to promote major firms in the main sectors of the industry. It was unashamedly nationalistic and based on the judgement that assistance and protection can help (or even ensure) the development of efficient firms in a new activity. Their policies, notably the extensive R&D assistance and the natonalistic purchasing, were well suited to this approach. It is a high-risk policy in that, if their lead firms do not compete successfully, their sub-contractors will also fail.

The Norwegians have so much oil that the terms of trade make exporting offshore technology difficult. However, they wish to be competitive in international markets in the high-technology, research-based activities, and their own experience in this area may give them some comparative advantage. The Norwegian case shows a major effort to move very rapidly through the initial learning and adjustment to continuous dynamic change in the technologically advanced sectors.

The Dutch have given substantial assistance to basic research and training but basically rely heavily upon free competition. This policy seemed successful in the early years when firms such as Heerema and HBG entered the market but more recently it has come under pressure. After 1975 considerable help was given to ailing firms, particularly in the shipbuilding sector, and the efforts to promote innovative developments in co-operation with lead firms reflects an appreciation of the competitive importance of technological innovation.

Three general lessons emerge from the experience of these four European countries. First, in an industry as complex as this, both lead firms and sub-contractors are important. However, full and fair opportunity policies have been more successful in assisting sub-contractors than in promoting major lead firms. Although very important, the capacity of sub-contractors to export depends on both close relations with lead firms and, except in high-technology activities, on open competition in the foreign markets. Promoting sub-contractors does not therefore necessarily lead to success in international markets. Second, an early start is important. Where projects are large and technical, and progress incremental, advantage goes to firms with experience and expertise acquired cumulatively. Lead firms (both contractors and technological leaders) take time to develop. The third lesson is the difficulty of dealing with an industry where risk plays so big a part. To be successful, a firm needs to be financially strong and able to take a long-term strategic view of its own development. Protection may help, particularly in promoting long-term confidence, but nationalistic policies, such as that pursued by the French, though they may bring great success, are in themselves very risky. Broadly competitive policies are much less risky and make the learning process easier, but the degree of success possible is not as great and some Governments may, with hindsight, regret what they see as lost opportunities.

Notes

1. The first platform used in the North Sea only weighed 320 tons, as compared with the great oil platforms now in use in the northern basin which reached over 500 ft. in height and are often over 40,000 tons.
2. After serious cost escalation and delayed completions, the oil companies increased their involvement in the big development projects. The big oil operators became more important as clients and the big projects were divided between major contracting firms.
3. The shipbuilding industry, having failed to win orders up to 1966, did relatively badly thereafter. After nationalisation, little was done until a new effort in 1981/2. The renewed efforts met with serious productivity, efficiency and labour difficulties and it did not prove possible to sell the newly designed semi-submersible. One of the major offshore yards, Scott Lithgow, has now been sold to Trafalgar House, and another, Cammell Laird, is in serious difficulties. Trafalgar House has, since the late 1970s, built up a substantial position in fabrication and it recently took over major facilities from the British Steel Corporation (RDL). In shipping, the major firms made some attempt to develop the offshore side of the business, notably P&O and the Ben Line. The latter has been fairly successful in drilling on a relatively small scale; P&O were more ambitious in the early stages but have in part withdrawn. In both shipbuilding and shipping, some small firms had considerable success.
4. The National Enterprise Board was at that time the state investment company; it supported both risky (often new) technology, enterprises and firms whose survival or expansion it thought important.
5. Reference to Schlumberger in *The Economist*, 8 January 1983, pp. 54–5 and 16 April 1983, pp. 80–1.

6. The CEPM was instrumental in founding GERTH (Groupement Européen de Recherche Téchnologique sur les Hydrocarbures), a 'technology club' which, in exchange for some pooling of R&D with other EEC countries, has assisted the French industry in obtaining substantial financial contributions from the EEC.
7. De Gusto has a considerable reputation in drilling rigs and continued as an independent firm after the shipyard was merged into the IHC group. It also had rigs built on behalf of clients in Finland and Japan.
8. IRO is the Netherlands Industrial Council for Oceanology; it publishes annually a major catalogue which gives considerable details about the members' activities.
9. J. H. B. Huijskens of IRO in *Focus on Holland*, 'The Netherlands Offshore Industry', August 1983. When pressed to give a rough approximation, he estimated the offshore turnover in Holland at some £500–700m and the number of employees between 10 and 15,000.
10. Ibid.
11. P. J. van Erven Dorens and Cor de Feyter, *The Netherlands Offshore Industry: Developments and Prospects, 1984*, written at the request of the Netherlands Industrial Council for Oceanology (IRO), pp. 61–2.
12. Ibid., p. 62.
13. This statement of Dutch policy comes from the Minister of Economic Affairs in *Focus on Holland*, 'The Netherlands Offshore Industry', August 1983.

Bibliography

Cook, P. L. and Surrey, A. J. (1983), 'Government Policy for the Offshore Supplies Industry', SPRU Occasional Paper Series, No. 31, October.

Delacour, J. (1981), 'Prospective des actions techniques conduites dans le cadre du CEPM...', *Guide Offshore*.

Department of Energy Blue Books (1975–6), *Offshore oil and gas: a summary of orders placed by operators of oil and gas fields on the UK Continental Shelf*, London, HMSO.

Department of Energy Brown Books, *Development of the oil and gas resources of the United Kingdom*, London, HMSO, annual since 1977.

Gazette Notice (1985) for the 9th Round of Licensing—Petroleum Production Regulations.

IMEG (1972), (International Management and Engineering Group of Britain Ltd.), *Study of potential benefits to British industry from offshore oil and gas developments*, London, HMSO.

MOE Report (1980), *Cost Study—Norwegian Continental Shelf*, Report of Steering Group Submitted to the Ministry of Petroleum and Energy, Oslo, April.

NEDO (1985), *Offshore Supplies*. A report by the joint offshore Committee of the Engineering and Process Plant EDCs.

Noroil (1982). *Survey of the French offshore oil industry*, August.
Peat, Marwick and Mitchell (1981), *Analyse des marchés det des grandes tendances technologiques dans la domaine para-pétrolier offshore*, Ministère de l'Industrie, Paris, CODIS.
Rödseth, T. (1981), 'Oil production and Norwegian resources' in *Kvartals-Skrift*, Special Issue on Oil and Economics, Bergen, Bergen Bank.
Saumon, D. and Puiseux L. (1977), 'Actors and Decisions in French Energy Policy', in *The Energy Syndrome: Comparing National Responses to the Energy Crisis* (ed.) L. Lindberg, New York, Lexington Books.
Scottish Development Agency (1985), *The International Offshore Market*.
Smith, N. (1978), 'The Offshore Supplies Industry—the British experience in its wider context', *The Business Economist*, Autumn.

Glossary

DRILLING

Jack-up rigs
Platforms carrying drilling equipment which can be floated from place to place and, once in position for drilling, can be jacked up on legs which go through the platform to the sea-bed. Such platforms are generally used in water depths of 70–300 ft. (e.g. the shallower parts of the North Sea).

Semi-submersible rigs
Drilling equipment is mounted on a platform which achieves its stability from pontoons submerged beneath the level of wave motion. The 'legs' which support the platform on the pontoons make photographs of semi-submersibles look very similar to fixed platforms. They are held in place by anchors and/or dynamic positioning, using their own power.

Semi-submersibles are now used for heavy-lift 'barges' and for major support/security/fire-fighting vessels.

Well-logging
The measurement and recording of the characterisation of a hole, particularly the geology of the structures through which it passes. Elaborate and highly sophisticated down hole instruments and recording equipment are now used to aid petroleum engineers and geologists to interpret the data.

PRODUCTION PLATFORM

Deck
The part of the production platform above water level which holds the equipment and accommodation modules. The deck may simply support modules, but it is now usual for the deck and modules to be designed as an integrated structure.

Jacket

This term is used for the tubular steel structures which, piled to the sea-bed, constitute the 'legs' of a steel platform. They surround the riser (a pipe connecting the well-head to the process equipment) and get their name from the function. These structures can be very large—the biggest, Magnus, weighing about 40,000 tons—and nearly 1,000 ft. in length.

Modules

The units in which the required equipment, service and accommodation modules are constructed and transported for assembly with the deck sections. This is mainly done offshore, although for small platforms the whole deck section may be assembled before being floated out.

8 Conclusions: Technology gap or management gap?

Margaret Sharp

The purpose of this study has been to examine the process of economic change in Europe by looking in detail at what is happening in six sectors, all of which are in the throes of rapid change. In two of these sectors, telecommunications and machine tools, well-established industries are observed adjusting to major changes in their technological base; in another two, videotex and computer aided design, we observe new activities developing around some of the products emerging from new technologies. In another—biotechnology—an old technology is being transformed by scientific discovery and we observe the early, tentative development of new products and the beginnings of a much wider impact across many sectors. Finally, in the offshore supplies industry, the impetus for change comes not from science and technology, but from the search for new sources of energy and in particular from the exploration and opening up of the gas and oilfields of the North Sea.

These six sector studies present, in effect, six snapshots of the process of change. Each focuses on a different subject, from a slightly different angle, and each in its own way has a different story to tell. And just as six snapshots do not make a movie, so these six sector studies do not, and cannot, present a comprehensive picture of the process of change. They do, however, suggest that there are a number of features which are of interest when change is particularly rapid. The purpose of this chapter is to identify some of these common strands, to consider how far they contribute to a wider understanding of the process of change and what policy conclusions, if any, they point towards.

The six case studies

Before embarking on a discussion of these common themes, it might be useful to recapitulate some of the main conclusions derived from each chapter.

COMPUTER AIDED DESIGN

Computer aided design (CAD) is an example of a new activity which has emerged as a new product of microelectronics. It originally developed as a highly specialist activity associated with the aerospace industry, with the United States leading the way because of major defence contracts, but with France and Britain both developing appreciable specialist facilities in the 1960s. Popularisation of the techniques waited on the fall in computing costs that came in the early 1970s, and the small, new American firms in particular forged ahead on the basis of 'packaging' the techniques for drawing offices and design shops. These small firms, established at this time as spin-offs from the American defence contracting projects, rapidly made their mark and by the end of the 1970s had become 'big league' suppliers of standard CAD equipment for both the United States and European markets, tying their customers to them by making their equipment and programing incompatible for wider use. By contrast, the French and British firms remained in the 'little league', finding niches for themselves either at the highly specialised end of the market or as suppliers of simple, but very limited systems.

From a European point of view, the interesting question explored in this study is why, in spite of their early start, European firms failed to make 'big league' status. The failure seems to have been primarily one of marketing—the European firms failed to seize the opportunities of rapidly falling computer costs to package and sell to potential customers 'user friendly' systems with popular appeal. But there were also other factors: Europe lagged behind the United States in the use and diffusion of minicomputers in the early 1970s, so the popularising technology was not so readily available, and the higher status of the draughtsman in European drawing offices made for difficulties in implementation. There also seems to have been a tendency for the French and British teams who developed CAD systems in the 1960s to want to remain at the high-tech end of the spectrum, and not to involve themselves in the less sophisticated popular techniques.

The continuing reduction in computing costs and the shift to the 32-bit computer presented the European firms in the early 1980s with an opportunity to break back into the field, but again it would appear that this has been an opportunity lost. It is arguable, Senker and Arnold suggest, that a co-ordinated European effort could have put at least one European firm into a big league position. Instead, the diverse and diffuse policies pursued by the different European

Governments have led to some of the best of European technology being bought up by American multinationals, with Europe left firmly in the 'little league' category.

ADVANCED MACHINE TOOLS AND ROBOTICS

Advanced machine tools and robotics represent two further 'islands of automation' which are gradually expanding within the manufacturing process towards a situation of computer integrated manfacture (CIM). Like CAD, computer numerically controlled machine tools (CNC machine tools) developed in the United States as a by-product of attempts to improve the precision and consistency of work in the aerospace industries and, as with CAD, Europe's first experiments with these machines also came in the aerospace industries, with both the United Kingdom and France developing 'cells' of highly sophisticated knowledge and application. The coming of microelectronics likewise brought the costs of the 'electronic' input tumbling down to a level where the technique could be 'popularised' with the introduction of general-purpose machining centres, while at the same time it enabled the 'robot' to be developed from its 1960s version (expensive, dedicated and somewhat inflexible) to a far more flexible (programmable) general purpose tool. Unlike CAD, it was the Japanese, not the Americans, who developed and popularised both production tools.

Of European manufacturers, only the Italians recognised the market potential for the relatively cheap all-purpose machining centre, and have succeeded in retaining their market share beside the increasing penetration of the Japanese. The other European success story, Sweden, went in the other direction, specialising in the production of robots and highly sophisticated applications of electronics. Meanwhile, the German, French and British industries foundered under the combined weight of recession and Japanese import penetration.

The industry now emerging from these strains is both chastened and changed. There have been a number of important developments. First, it has become clear that the future lies with electronics: the successful firms have all formed close links with suppliers of electronic controls. Second, having traditionally been highly integrated concerns (making their own precision parts because they wished to maintain control over quality), many successful companies are now dis-integrated concerns, buying in, often to computer specification, parts made by outside component suppliers. Third, the new techniques have revolutionised the economics of batch production, making feasible a flexibility in production hitherto unknown.

Interestingly, this trend has reversed the traditional customisation in the machine tool industry itself—programmable general-purpose machines now substitute for dedicated machines. The result is that while CNC machines and robotics have overturned the traditional economics of batch production for machinery users (i.e. customers), they have actually made economies of scale *more* important in the machine tool industry itself.

The European industry is still in the process of assimilating these changes. For the West German industry, whose strength was built on craftsmanship in precision engineering, the recognition that electronic skills are now more important than metalworking skills has come hard, but there are signs that it is now regaining its traditional strength as a bulwark of the German mechanical engineering industry. The British industry has been decimated by a mixture of recession and incompetence, both in the industry itself and on the part of government. Current policies favour teaming up with Japanese producers on the grounds that 'if you can't beat them, join them', but such policies are regarded with great suspicion elsewhere in Europe. The French industry has never been a strong one and in spite of a substantial injection of government funding remains relatively weak. Only the Swedish and Italian industries remain relatively unscathed through the process of adjustment.

TELECOMMUNICATIONS

The reason for studying telecommunications as a new industrial activity is because, after fifty years of relative technological stagnation, developments in technology within the last decade are transforming the industry. In the first place, having for many years been concerned exclusively with voice telephony, the industry is increasingly involved in data communication with the demands on the network from this source growing very fast indeed. Add to this developments in cable technology, fibre optics, satellite communications and home computers, and the range of potential services which can be provided by the same facilities multiplies. On the equipment side, because the network is so complex, change is by necessity piecemeal—a gradual adaptation to new technical opportunities. The current generation of digital electronic exchanges, for example, has taken some ten years to develop and will not be fully installed until the mid-1990s. The result is considerable tension between this gradual pattern of network adaptation and the burgeoning of new activities which put new demands on the network. It is this tension which is the origin of many of the problems the sector currently confronts.

Attention in this study focuses on the process of transformation: how an industry that had seen little technological change for some fifty years adapts to such a technological upheaval and to a period of fast and continuing innovation. On the one hand, there is the varying experience of the major European public utilities: from the French attempt to use the position of telecommunications industry at the centre of the *filière électronique* as a means of establishing and reinforcing a French presence over the whole *filière*, to the British moves to dismantle the public utility altogether. On the other, is the challenge to the established equipment manufacturers and the distintegration of their cosy relationships with the various public utilities. Overshadowing all developments are the repercussions in Europe from the dismantling of the AT&T monopoly position in the United States and the potential entry of both AT&T and IBM into the European market. We may observe the changes that have so far taken place—the strategies adopted by both firms and Governments—aware, however, that the drama is as yet only halfway through.

VIDEOTEX

Videotex is an example of one of the burgeoning new activities available within the telcommunications sector. It provides an interactive link between television, telephone and computer so that the user can key, via television set and telephone line, into large data bases and/or other computer systems. Its technology, even when first introduced in the early 1970s, was relatively simple, but had the advantage of offering interconnectability via the existing telephone cable network (rather than having to create a whole new cable network as for cable TV). But the main advantage to its pioneers, British Telecom with their Prestel service, was a marketing one— to increase revenues by using surplus capacity in the telephone system. In spite of this, British Telecom badly misjudged its pricing and marketing policies and was forced to redefine its strategy within two years of opening the service. The lessons from some of these mistakes were learned by the French and the Germans as second-round innovators, and the German service, in particular, is closely geared to market needs. The French, somewhat typically, tried to use the opportunity to leap-frog the French system into a commanding position in world videotex, but this attempt to run before it could walk rapidly came to grief on both political and economic grounds.

The question remains whether, with the rapid developments in cable technology which open up the opportunity of a far more comprehensive interactive service, there is any future for the more limited facilities offered via these PTT-dominated systems. There are positive advantages in using the existing telephone network in terms of economies of scale and inter-connectability, while the new fibre optic cables planned for many cable networks still have to prove their economic viability. It has yet to be seen, however, whether in the long run these advantages offered by the public monopoly PTTs in Europe outweigh the greater flexibility (to market need) of the private systems. To date, the piecemeal development of American experiments indicates the advantage of some unifying influence even if, as in Canada, it is adherence to a unified standard. The advent of the home computer and the desire for a means of communication via computer seems likely to give these systems a new lease of life.

BIOTECHNOLOGY

Unlike the activities discussed so far, where developments in one way or another have been seminally influenced by microelectronics, biotechnology is a distinct technology—or, to be more exact, a cluster of technologies—and one that in itself will have widespread repercussions upon other industrial areas. Just as the semiconductor introduced a new era for electronics, so genetic engineering has brought a new era into biotechnology.

This new biotechnology is still in its infancy—many of the techniques being used are barely a decade old even as scientific discoveries, let alone as industrial activities. A study of biotechnology in its present phase of development therefore encapsulates many of the issues which arise in the translation of scientific activity from academic research laboratory to commercial activity—issues concerning the relationship between science and technology, the uncertainties which surround scale-up from laboratory to factory and the development of new products and new markets: issues of finance and access to information, of the role of large and small firms, of Governments, universities and research institutes.

Three main conclusions emerge from this study. First, with commercial development still limited to the high value added areas of pharmaceuticals and fine chemicals, advantage still lies at root in the research base, and the varying capabilities of European countries still reflect this competence, with the United Kingdom and West Germany emerging as the strongest European contenders. Second, in spite of widespread Government programmes of support and intervention, the process of commercialisation is dominated by industry itself and, in the case of Europe, by the large chemical/ pharmaceutical multinationals. This puts considerable emphasis on the linkage mechanisms between the research base and industry and Europe's mechanisms are by no means as clear-cut as the market-based small firm intermediary of the United States or the collaborative mechanisms of Japan. Third, uncertainty as to scientific developments still dominates the decision-making process—time horizons are long and pay-offs uncertain. In these circumstances there is a danger that the oligopolistic structure of the European industry may tend to encourage a strategy of establishing capabilities but then 'waiting' to see what competitors do. In contrast, in the United States, the use of the small research firm provides a hedge against technological uncertainty (and a stimulus to competition) and amazing transparency of action. In Japan, the extensive collaborative but corporatist framework gives the reassurance of mutual support in decision-making. Overall, in a field as young and diverse as biotechnology, there are many opportunities to be seized. The study concludes that Europe may best exploit these by building on comparative advantage, buying in knowledge and technique where necessary, rather than by contemplating *ad nauseam* its position in league tables with Japan and the United States.

OFFSHORE SUPPLIES

The offshore supplies industry is dominated by three related characteristics. First, it is concerned with large, complex, one-off projects, some costing well over £1,000m. Second, the rate of technical innovation is high, but continuous and incremental, as the industry tackles wells in ever deeper and more difficult areas of the North Sea. Third, handling these large projects in this situation of continuous, fast, incremental change requires a large number of experienced personnel, most of whom are provided by sub-contractors, with the major management and contracting engineers at the apex of a very large pyramid of sub-contractors. Since knowing how to do things and having experience of doing them are of vital importance to successful project management, experience itself becomes a scarce resource and leads to the 'spin-off' phenomenon when individuals break away from their old firms and set themselves up in business. The industry, in effect, is built upon layers of sub-contracting, the whole forming a complex network with links into many other industries. Where, as an industrial activity, the industry begins and ends, is a moot point.

Development of the industry in Europe is primarily associated with exploitation of North Sea gas and oil off Holland, Britain and Norway. The British and Norwegian Governments both give substantial protection to their own industries—Norway more than Britain since the British 'full and fair opportunities' policy treats resident multinationals as British firms. But protection in Britain came relatively late in relation to development, and the main areas of process engineering and drilling equipment are dominated by experienced foreign firms which came in early and derived further advantage from North Sea experience. As a consequence, while British companies are well represented in the industry, they tend to occupy specialist niches as sub-contractors rather than to dominate mainstream activities. Norway, by contrast, pursues a more protectionist policy and virtually excludes most foreign firms, except in joint ventures. Holland had from the start a very open market; it developed successfully at an early stage a sophisticated shipbuilding industry and much dredging and barge operation expertise, mainly on the basis of gasfield experience. The firm Heerema dominates the heavy-lift market world-wide but it is now as much a multinational as a Dutch firm.

Perhaps the most interesting experience is that of France. Although there is no oil or gas off the coast of France, the two state owned oil companies (Total and Elf Aquitaine) were encouraged to acquire and develop oil reserves, initially off Africa and in the Middle East, and subsequently in the North Sea (and now China). Using their purchasing power as major contractors, these firms have consistently since the early 1960s set out to build up and promote a network of firms to form a French offshore supplies industry. This policy has been reinforced by the exclusion of multinationals and a programme of substantial, publicly funded research. Together, these measures have combined to nurture a strong French presence in this industry.

Some common themes

In the introduction to this chapter it was suggested that there were a number of common themes emerging from these six sector studies which might merit further exploration. This section develops a number of these ideas.

(i) EVOLUTION OR REVOLUTION?

The central queston underlying these six sector studies is the question of *how* new industrial activities are emerging within the economies of Western Europe. The straightforward answer to this question is to say that, on the basis of the six sector studies presented here, these new activities are emerging largely from within existing industries—even indeed from existing firms within those industries. The crucial decisions about new activities are the decisions taken by firms—decisions as to when to pick up a new idea, how much to invest in it and at what stage of development. In this sense the process of change is evolutionary—new industrial activities emerge from the body of old industrial activities, the decisions are incremental as firms adjust their product/process mix to opportunities which present themselves, and, as this happens, so firms progressively redefine the nature and boundaries of the industry itself. New activities, new industries, evolve from existing industries; they do not suddenly burst on the scene in complete and finished form. As Lesley Cook warns in her study of the offshore industry, 'newness is a matter of degree'.

The evolutionary nature of new industrial activities in turn reflects the evolutionary process of much innovation. There has been considerable debate on the question of how far the governing influence over innovation is market demand or changes in technology. The 'demand pull' proponents maintain that the main impetus to innovation derives from scarcity and shortage stimulating new production methods; the 'technology push' school maintain that the main influence derives from scientific discovery which progressively pushes forward the bounds of production feasibility. As Mowery and Rosenberg (1979) document in their survey of the debate, successful innovation requires *both* technology push *and* demand pull: there are major weaknesses in the conceptual framework of many innovation studies because they have failed to recognise the interactive nature of innovation. And it is the interactive nature of innovation that helps to explain its predominantly evolutionary character.

The studies presented here certainly reinforce the continuous, interactive view of innovation. Take the offshore supplies industry, chosen for study precisely because it was 'demand led'. Yet the need to cope with ever deeper and more difficult offshore conditions has demanded the constant adaptation of technology—indeed the offshore industry is a prime example of an industry where there are 'innumerable ways in which changes, sometimes very small changes in production technology, are continually altering the potential costs of different lines of activity' (Rosenberg, 1982, p. 231). By way of contrast, let us take the one activity amongst the six which might be described as most nearly pure 'technology push'—videotex. This amply illustrates the need for an interactive approach, for it was precisely because the PTTs failed to identify a clear market for it that it has so nearly failed as a new industrial activity.

The concept of trajectories is useful in this context. A trajectory describes the path taken through time by a technology subject to continuous incremental innovation (Dosi, 1983). A new technology very often, and certainly in the cases we have been studying, is subject to continuous improvement over a number of years. Firms which develop the capability to make these continuous improvements, that is to move along the trajectory, are often the most successful. As well as continuities, there are discontinuities. Major new technical or marketing innovations present such discontinuities. A discontinuity halts progress along the existing trajectory but simultaneously opens up a new one. In assimilating major technological change, a firm in effect changes gear and shifts to a new trajectory. In this sense, the discontinuity may be regarded as a revolution. Whereas evolution, development along the trajectory, is an everyday occurrence, revolutions are quite rare.

There are two examples of 'revolutions' in the studies presented in this book. One is genetic engineering, discussed at length in the biotechnology study. Although its exploitation is dependent upon teaming the new techniques with existing process technologies, genetic engineering represents a major breakthrough of technique and holds out the potential for major developments in the worlds of medicine and food production. As yet we have experienced very little of its impact. A generation earlier, similar breakthroughs were occurring in the application of solid state physics to the development first of the transistor and subsequently of the semiconductor —a revolution which was to bring upon us all the deluge of developments in microelectronics. The story of CAD documents one of the mainstream developments that has come in the wake of this 'revolution': in videotex, telecommunications and advanced machine tools, the focus is upon applications rather than mainstream developments. While the process of assimilation will inevitably dampen down and mute the 'shock wave' from discontinuous scientific discovery, it does not, nevertheless, totally eliminate it. Many of the 'new' technologies' which are currently being introduced stem from the discontinuity in microelectronics—a host of sectors are shifting from old to new trajectories.

(ii) NEW FIRMS OR OLD FIRMS?

A subsidiary question in this debate is whether the new activities are pursued by existing firms or new firms. The evidence that rapid change can be undertaken by old firms is surprisingly strong, although it is sometimes difficult for a firm which has been using one technology to change substantially. (The problems relate to management and attitudes as well as technology *per se*.) This is well demonstrated in the case of CNC machines where quite large numbers of firms failed to make the jump. It is also demonstrated in the case of offshore where firms doing work in similar areas and with many of the necessary skills (e.g. in the shipbuilding industry) were unable to meet the required standards of reliability and speed for offshore work.

Discontinuities often give rise to opportunities for new entry. Shifting to a new trajectory of development means that old skills and old experience are of less value, while new skills in the new areas of development are scarce and command a high premium. Those possessing these new skills are often tempted to break away from established firms and set up on their own, which helps to explain the flowering of small innovative firms at the start of a new trajectory—witness the experience in CAD and biotechnology. But the other interesting feature is that it also opens up opportunities for new entry from firms outside the industry (as has been happening in telecommunications).

A contrast between European experience and that in the United States suggests that by and large in Europe new activities have emerged from existing firms rather than from new entrants, although there has been some sideways movement amongst large firms (for example, Olivetti moved into electronic machine tools; Matra, the French defence/space firm, moved into telecommunications equipment). Most noticeable, however, has been the absence of the small, new, innovative firm in both microelectronics and biotechnology and, perhaps most of all, the relative rigidity of ranking amongst the major firms. Table 8.1 illustrates the changing fortunes of the leading American electronics firms between 1955 and 1975, with only one firm, RCA, left among the top ten after twenty years. An equivalent table for Europe would have found most of the large firms—Siemens, Philips, Thomson—still in place!

It is unclear how far this relative rigidity in ranking, and above all the lack of new entry, really reflects the rigidities of the European markets. After all, a similar rigidity in ranking would be noticeable

Table 8.1 Top ranking American electronics component producers, 1955 and 1975

Rank by sales volume

1955 Valve technology	1975 Integrated circuit technology
1 RCA	1 TI
2 Sylvania	2 Fairchild
3 GE	3 National Semiconductors
4 Raytheon	4 Intel
5 Westinghouse	5 Motorola
6 Amperex	6 Rockwell
7 National Video	7 GI
8 Ranland	8 RCA
9 Eimac	9 Philips
10 Landsdale Tube	10 American Micro

Source: Barker (1984).

among Japanese firms, although few would accuse them of lack of innovation. In part, the contrast reflects institutional differences. Europe lacks the private universities of the United States, with their traditions of contract research, and it lacks the venture capital markets which have helped the academic entrepreneur start up on his own. But to too great a degree the ambition of small European firms seems to be a (profitable) take-over by a large multinational rather than the mounting of a challenge to the market placing of that multinational.

(iii) THE PERVASIVE INFLUENCE OF MICROELECTRONICS

A factor which comes through very clearly from these six case studies is the current pervasive influence of developments in microelectronics. In four of the six studies—CAD, videotex, telecommunications and machine tools—it provides the main impetus for change. In one sense the finding is trivial. We cannot pretend that the selection of sector studies is random: one of the underlying thoughts was to explore the impact of microelectronics outside its mainstream activities (computers, semiconductors) and we have found precisely what we set out to find! In another sense, however, the finding is less trivial. Had we set out to explore the development of, say, the new ceramics or lasers, we would have found the same —that developments in microelectronics were enabling improvements to be introduced which otherwise might not have taken place, or might have taken a great deal longer to emerge.

The term 'generic technology' is sometimes applied to those technologies which spill over widely from their mainstream activities to affect many other areas of production and development. This contrasts with 'specific technologies' whose impact is limited to the specific industry concerned—for example, continuous casting is a specific technology in the steel industry, whereas microelectronics and biotechnology are both generic technologies whose impact is widespread. In the previous section we identified these two technologies as incorporating important discontinuities in their current phase of development, and introduced the notion of the new trajectory of development. The combination of these two characteristics—the generic technology and the discontinuity—is an extremely powerful combination. In the past, such combinations have been associated with technologies such as steam or electricity; in the twentieth century with petrochemicals and the internal combustion engine. Each is not only a major technological advance in its own right (and therefore attracts new investment and equipment into its mainstream activity), but spills over to stimulate new products and new processes elsewhere (for example for petrochemicals, man-made fibres and plastics), bringing with it a secondary round of innovation and investment as new trajectories are explored and developed.

The concept of the locomotive technology with its associated clusterings of innovation derives from the work of Schumpeter and it is worth spending a little time considering more closely his ideas, and those that derive from them.

Schumpeter's basic model, expounded in his Theory of Economic Development (1912), is based on a three-stage process:

(i) an exogenous but discontinuous flow of basic inventions related to developments in science;
(ii) the exploitation of these inventions by a group of spirited entrepreneurs who are prepared to take high risks in return for high rewards;
(iii) the swarming of secondary innovators who are attracted by and compete away the monopoly profits of the early innovators.

Later in his life, Schumpeter incorporated an endogenous element into the development of science and technology: with R&D activities increasingly under the control of large firms, many of the major scientific breakthroughs would derive from the laboratories of these organisations and the process becomes self-reinforcing (Schumpeter, 1939). He also suggested that an explanation of the Kondratiev long cycle might be linked to the exploitation of basic inventions and their associated swarming process, but it was left to Kuznets (1940) to point out that this argument could only explain the major fluctuations of Kondratiev dimensions if there were a clustering of inventions in the decades associated with the trough of depression, their exploitation therefore being likewise concentrated in the upswing period.

In their recent book *Unemployment and Technical Innovation*, Freeman, Clark and Soete (1982), dismiss Mensch's (1975) attempt to establish a theory of clustering of innovation as a result of depression-induced acceleration in the gestation period between basic inventions and basic innovation. In its place they put emphasis on the diffusion process, arguing that the exploitation of an invention may be delayed for a decade or more until profitability has been clearly established and/or other basic facilitating and organisational changes are made.

> Once the swarming starts it has powerful multiplier effects in generating additional demands on the economy for capital goods (of new and old types), for materials, components, distribution facilities, and of course, for labour. This, in turn, induces a further wave of process and application innovations which give rise to expansionary effects in the economy as a whole. [Freeman, Clark and Soete, 1982, p. 65.]

The whole process of diffusion involves not merely imitation of the original invention, but frequently a string of further innovations—large and small—as an increasing number of firms begin to learn the new technology and to use it to gain advantage over their competitors. This wave of secondary innovation in effect creates a whole series of new trajectories.

It is not difficult to relate this analysis to the development of microelectronics. The major technological breakthrough came with the transistor and the integrated circuit, and the clustering of innovations around these breakthroughs led in turn to major cost reductions. There has since been a swarming of new firms, each jockeying for position with each other, pushing out the old market giants, then finding themselves subsequently challenged, sometimes by other new firms, sometimes by the old market leaders seeking to re-establish themselves.

Viewed in this vein, many of the activities studied in this book can be seen as being caught in the wider nexus of developments in microelectronics, and still being affected by the 'shock waves' caused by the discontinuities of the 1960s and the subsequent fast development of the mainstream activities. CAD emerged as a result of the rapid fall in computing costs and the increase in computer capabilities that came with the microprocessor and the minicomputer; videotex was seen as a method of harnessing the data processing capabilities of large computers via the telephone service into the home; telecommunications and machine tools both echo the 'discontinuous' impact of microelectronics which has shaken them out of a long period of relative technological stagnation and both face a period of turmoil and uncertainty as the new (microelectronics-dominated) technology is assimilated.

(iv) THE *FILIÈRE* AND THE PRODUCT CYCLE

The concept of the *filière* has already been introduced in the chapter on videotex but its application goes considerably wider. The nexus of relationships around microelectronics might alternatively be called the *filière électronique*—the *macrofilière* associated with the electronics industry and embracing the upstream and the downstream relationships within the nexus. As was made clear in the videotex chapter, some French economists (Malsot, 1980 and Mistral, 1980) take the linkage further and have developed behavioural relationships based on the *filière*, suggesting that in the early stages of the development of a new technology (i.e. close in time to the discontinuity point), firms controlling the upstream technology (e.g. semiconductors) can dominate the whole *filière*, but that control moves downstream towards the consumer as technology matures.

This argument has enormously important policy implications, for it suggests that the successful *application* of new technologies downstream depends upon strong links with upstream developments (for example the development of the computer industry in relation to semiconductors; telecommunications or CNC machine tools in relation to computers). This in turn has been used to support the seemingly chauvinistic argument that strength in one part of the industry depends upon a chain of strong firms, upstream and downstream, which rapidly turns into an argument in favour of the promotion of 'national champions' in what are seen to be 'strategic industries'.

This French approach is seen at its best in the offshore supplies industry. Here an early (circa 1960) decision was made to establish a substantial French presence in the industry; the two nationalised state oil firms, being in the pivotal position within the *filière*, were used to foster through their purchasing policies a chain of French firms supplying the needs of the industry; and the whole was backed up by a major government financed R&D programme. Through this programme of what amounts to infant industry protection, France has built up an industry which is now competitive with the international giants. By contrast, the British offshore industry, which was (rather later) given an element of protection, but where policy lacked the highly nationalistic orientation of the French, has had difficulty in establishing itself in the mainstream activities of the industry.

The studies of telecommunications and videotex chronicle another, but less successful, attempt by the French Government to use a pivotal organisation (in this case the DGT) to dominate the whole *filière* and exert its leverage to raise the competitiveness of the French (electronics) industry. More broadly, while this approach has worked well for the French in some sectors (nuclear, space, for example), they have consistently had trouble in trying to organise the less concentrated electronics sector along similar lines.

The debate about *filières* goes wider than specific sectors or groupings of sectors to the more general debate on whether there are some industries which might be 'strategic' industries. Horn, Klodt and Saunders view with scepticism the arguments that have been put forward to suggest that the machine tool industry occupies a 'special place' in the economy, but Arnold and Senker argue that Europe's failure to consolidate its strength in CAD has made the European industry unduly vulnerable to the whims of American trade policy, particularly to strategic embargoes on high-technology equipment. They also argue that failure to develop an indigenous industry means a failure to build up a cadre of technicians able to use these skills to maximum benefit. Cook makes it clear that the French policy in relation to the offshore supplies industry was not costless: it was a high-risk policy pursued (consistently) over a large number of years. Perhaps the most salutary note of warning comes from Dang Nguyen in the telecommunications chapter. Here the *filière*, in the form of the chain of equipment suppliers to the telecommunications authorities, had been protected and cosseted over a matter of fifty years. Far from using this strong home base to develop an internationally competitive industry, the firms involved had, given the chance, exploited their home markets, behaved opportunistically towards their PTTs and failed to develop internationally competitive products. The answer is perhaps that infant industry protection should not last the length of the product cycle!

At the opposite end of the spectrum to the French attempts to control the *filière* by promoting French firms at strategic points in the chain, is the Swedish approach which recognises the impossibility for a country of Sweden's size of seeking to develop competence across the whole range of new technologies. Instead, the policy has been deliberately to 'buy in' technological know-how via licensing and joint ventures to complement home-based skills. The relative success of Sweden in a number of areas of new technology—robotics, CAD/CAM and biotechnology—and the continuing ability of its major multinationals such as Volvo and Ericsson to hold their own in international markets, lend weight to Horn, Klodt and Saunders' scepticism about 'strategic' industries.

The key issue in the *filière* argument relates to knowledge and experience. Sweden's highly skilled and highly trained work-force is well able to assimilate foreign techniques and adapt them to Swedish needs. Even when the requisite knowledge and experience are not there, they may be bought in, provided there is (as in the offshore supplies) an active and likely international market for such skills. The case for developing core industry groupings rests on the absence of this market. It may not exist in the early stages of the development of a new technology: key skills and knowledge may reside with major firms which have consolidated their position through patents; major producers may have bought up scarce expertise at a monopoly price in order to preserve their market positions; or the developments may be so new that no market has yet established itself. Alternatively, there may be no effective *international* market, which is where the American embargo on high-technology exports makes its impact. But in the absence of such embargoes, it seems that it is only in the very early stages of the development of a new technology that knowledge and skills can be monopolised to such a degree that action to promote competence at several points in the *filière* is justified.

(v) THE SCIENCE-BASED INDUSTRY

One notable feature of the six sector studies is the degree to which new developments derive from scientific discovery. It is particularly apparent in biotechnology which is, in any case, still half-in, half-out of the the research laboratory; but most of the developments in microelectronics were/are contingent on developments in solid state physics and materials technology, and even the offshore supplies industry—in many senses the least scientifically orientated of the six—has been dependent upon developments in underwater surveying and marine oceanography.

This dependence upon science is no new phenomenon. Landes (1969) notes how industrialisation during the nineteenth century shifted from the 'tinkering' technology that marked the first industrial revolution to the science-based development of the chemical, steel and precision engineering industries at the end of the nineteenth century (and the concomitant relative decline in Britain's industrial stature *vis-à-vis* Germany and the United States). The importance of access to basic and applied science seems if anything to be increasing. A recent American report pointed to the shortening lag between basic research and technological application.

> The rapidity of scientific and technological advances links directly to the emergence of new research-based technologies. These are technologies that would not exist without basic research, for which the role of basic research was not to improve but to germinate. Thus . . . the intense industrial development of very large scale integrated circuits depends on a knitting of fundamental work done *almost in parallel* [italics added]—that work embracing materials and solid state science, a host of new spectroscopics, and atomic and molecular physics. [National Academy of Science, 1983.]

This shortening lag puts increasing emphasis on a country's scientific and technological infrastructure, composed of universities, polytechnics and research institutes.

This infrastructure has two functions. The first is the support and promotion of the basic sciences—research which may have no obvious immediate commercial pay-off, but which nevertheless serves to push forward the frontiers of knowledge and provides the font of ideas/discoveries which feeds technological innovation. Given the uncertainties of success, the variable time-lag between discovery and commercialisation, and the difficulty in establishing ownership over academic ideas, such basic research is normally treated as a public good and generally funded from the public purse.

The second but equally important function of the scientific infrastructure is the dissemination of scientific knowledge. This is achieved both by the training of new scientists and engineers and the incorporation in their training programmes of new techniques and new knowledge. One of the signs of a maturing technology is the extent to which it is incorporated into training programmes: vice versa, because these programmes train more and more people in the new techniques, so the need for very close links between mainstream technological development and successful application lessen, which helps to explain the *filière* relationships discussed in the last section.

Just as a strong scientific/technological infrastructure helps in the dissemination of new ideas, so too does it help to equip the labour force with skills necessary to make good use of these new science-based technologies. Swedish and German experience with CAD is interesting in this respect. Although neither country had had firms which had participated in the early development of the technology, the general level of skill training in both countries, particularly in engineering, was such that, once introduced, it was rapidly picked up and adapted to local working conditions. Countries such as the United Kingdom, which have a generally lower level of skill training, often find it more difficult to introduce new technologies because the work-force both lack the minimum requisite skills to make good use of them, and tend to be more suspicious of the new techniques. Time and again, lack of skills is identified as the major constraint on the development of new technologies. (See, for example, the recent report published jointly by the Policy Studies Institute and the Anglo-German Foundation—Northcott and Rogers, 1985.) With the increasing use of science-based technologies, this is a problem that is not going to disappear. Investment in education and training is an important part of the infrastructure for the modern industrialised economy.

If research, education and training are important, so too are the mechanisms for technology transfer—for linking the scientific infrastructure into society, and particularly into industry. The study of biotechnology illustrates the contrasting mechanisms that have evolved in the United States and in Japan. In the United States a new form of market intermediary has arisen—the start-up firm—often started by a group of academic scientists anxious to capitalise on their own skills and finding a ready source of funding—the venture capital market—only too willing to help them do so. By contrast, in Japan, the small firm hardly exists in this role. Research is undertaken in the research laboratories of firms and Government and the linkage is collaborative, grouping together research scientists from industry with academic and government research personnel, working to agreed targets on 'pre-competitive' research and development. When this is completed, the team disband and return to their 'home base' for the next competitive stage of development.

In Europe the linking mechanisms are by no means so clear-cut. In Germany and countries following the German university traditions (Switzerland, Netherlands, Scandinavia), the linkage between engineering faculties and industry has always been close, but this has not been so in the case of the pure sciences, and throughout Western Europe there is something of an élitist culture which for long has resisted the 'commercialisation' of science—a culture which is only just beginning to break down. As yet, too, the venture capital market in Western Europe is in its infancy—the only country which has really begun to follow the American path is the United Kingdom.

Meanwhile, the Japanese route of collaborative research is finding favour, particularly among Community bureaucrats in Brussels for whom the ordered, collaborative world of the Japanese has considerable attraction, not least because it provides an obvious niche for their services. Whether the competitive or collaborative route is better fitted to European culture and traditions is not clear. There are dangers that the collaborative route, without the Japanese spur to surpass the United States in competence, will be used as a mechanism to reinforce corporatist decision-making, rather than as a spur to competitiveness. Certainly, both national programmes to support microelectronics (for example Alvey in the United Kingdom) and the Community ESPRIT programme have tended to channel the bulk of funding towards the large, well-known firms. But they have also brought together firms and academic groups in a way hitherto unknown in Europe.

The history of CAD illustrates another facet of the process of commercialisation of science-based activities—namely the value of public purchasing contracts focusing the early stage of development. The demands of the USAF for a technique which could be used with numerically controlled machine tools to produce an engineering output with greater speed and accuracy gave the 'demand' for the technology, while the scientists at MIT and associated with the air frame companies of McDonnell Douglas and Boeing provided the 'supply'. Likewise, the use in France of 'demand-led' research contracts in the offshore industry, possibly because of the close relationship between the (part government, part industry-funded) research institutes and the main companies involved, has been a continuing source of strength to the French industry. These 'customer–contractor' relationships at an early stage of experimentation can help to target science-based developments while simultaneously providing the spur to commercialisation. However, as noted earlier, there are dangers of carrying this process over from infant to mature industries, as the history of telecommunications illustrates so well.

(vi) COPING WITH UNCERTAINTY

Discontinuities, new science-based industries, the shortening lag between basic research and technological application . . . all point to another important theme that has recurred in thse studies: the problem of coping with uncertainty.

It is a truism to suggest that capitalism is about risk, uncertainty and profit. Economists make a distinction between risk and uncertainty. In situations of risk, probabilities can be actuarially measured and therefore allowed for: uncertainty is the situation when probabilities are unknown or can only be guessed at. When change is continuous but incremental, experience lends expertise to guessing at probabilities and risks are often quantifiable. Thus, in the offshore industry, the risks of drilling a dry-hole are known and allowed for in exploration and development expenditures. In the telecommunications industry, where public monopolies had enjoyed fifty years of steadily increasing market penetration, investment decisions had become practically riskless—the equipment used was well-tried and reliable, and, even if not used immediately, new capacity would rapidly be absorbed by market growth.

The uncertainty is introduced by the discontinuities. In telecommunications there is new equipment with an uncertain period needed for development and system adaptation and with no record for reliability: the application of microelectronics to machine tools brings new entrants (the Japanese) into European markets: genetic engineering offers the possibility of a whole new range of drugs based upon therapeutic proteins. In all these cases past experience is of little value in predicting future product reliability or market behaviour.

The problems of coping with uncertainty are perhaps most vividly illustrated in the case of biotechnology where the discontinuity of genetic engineering has placed firms in the doubly uncertain position of not knowing whether the product they are trying to develop is (a) technologically feasible (i.e. can it actually be produced?) or (b) commercially viable (if produced, will there be a market for it?). One can regard such uncertainties as the stuff of entrepreneurial capitalism—high risks are rewarded by high profits and it is the gambling instinct of the entrepreneur that pushes forward the frontiers of production feasibility and translates scientific discoveries into commercial products. There is a sense in which the start-up biotechnology firms of the United States seem almost a pure embodiment of this process, backed up by a venture capital market that works on the assumption that only two out of every ten firms it backs will make good.

It is interesting, however, to look beyond the start-up firms to the strategy adopted by the major established corporations in order to gain an insight in to how the mature, multinational capitalism of the late twentieth century copes with such uncertainties. First comes the spreading of risk—most corporations will have a mixed portfolio of investments of which those involving biotechnology will constitute but a small part. Second comes the arm's length involvement, with no substantial in-house investment until they have narrowed down the area of uncertainty to risks they are prepared to take: witness how the major American corporations used the new biotechnology firms as a means of testing the technological feasibility of new genetic engineering techniques before building up in-house capabilities. A similar process of market research helps identify products with substantial market potential as distinct from those likely to have limited or overcrowded markets. Finally comes the defensive investment—the need to match capabilities of competitors and provide themselves with a 'window' on the technology irrespective of their own assessment of probabilities.

There comes a point, however, when entrepreneurs have to make subjective judgements on probabilities and take decisions in the light of what are regarded by the enterprise as acceptable risks. Looking across the sectors and countries studied in this book, it is at this point that there is fascinating interplay between culture and institutional mechanisms. For the Americans, the 'get up and go' culture of the pioneer still pervades and this is matched by the institution of the venture capital market, and considerable ease of entry and exit for the small firm. By contrast, the Japanese need to build up the slow but careful consensus on what is likely to happen in the future—hence the corporatist framework with its long-term strategic planning and mutually supportive decision-making procedures which short-circuit what might otherwise be risk aversion (because Japanese businessmen do not like to buck the trend of opinion).

Europe as a whole is marked by a greater degree of risk aversion than found in either the entrepreneurial capitalism of the United States or the corporate capitalism of Japan. With the exception of the United Kingdom, which has recently grown an active venture capital market, Europe wholly lacks the institutions necessary to back up entrepreneurial capitalism. Ergas (1984) finds both entry and exit mechanisms in Europe more cumbersome than those of the United States. Its banking sector has traditionally lent to big business but is conservative towards small business, and actively opposes the entry of American banks or financial institutions to fill the void. Given the banks' involvement with big business, the corporatist model might be more appropriate, but Europe generally lacks the consensus mechanisms built around long-term strategic planning which play such an important part in extending the corporate horizons in Japan. The French come closest to the Japanese model, but it is the state rather than the corporate sector which takes the lead. German and Swiss businessmen share with their Japanese counterparts a willingness to look to long-term product development, but do not favour joint decision procedures.

(vii) NATIONAL CULTURES AND PUBLIC POLICY
—HARES AND TORTOISES

It is always dangerous to try to ascribe national characteristics to observed behaviour; nevertheless, business schools are increasingly borrowing from social anthropologists and showing an interest in cultural factors which influence patterns of business behaviour. The previous section suggested that cultural factors might influence attitudes towards risk-taking, and that institutions, to a degree, mirror these attitudes. Continuing in this vein, there is another comparative across-country theme that comes through from the sectors—a theme which might be called the tortoise and the hare phenomenon.

France and Britain both have a tendency to leap on to the leading edge of new technologies, but to fail to consummate the advantage in more hum-drum, downstream applications. CAD provides an apt illustration. Both countries associated themselves early with developments in CAD, and built up specialist teams which have contributed substantively to some of its most sophisticated applications. But in spite of this advantage and the 'cells' of highly skilled personnel associated with it, in both countries it was left to the American multinationals to popularise the technique. Contrast this with the pattern of development in Sweden and Germany, neither of whom attempted these early leaps on to the frontier, but where there are highly skilled and trained work-forces in the engineering field, and who have come from behind, picking up and using the cheaper and widely applicable techniques developed by the Americans.

A not dissimilar story comes through in other sectors—telecommunications, biotechnology, even nuclear power (Surrey and Walker, 1981). For France, the 'leap to the frontier' is often associated with attempts to create and control the whole *filière*. This highly nationalistic approach to policy goes back a long way in French history, although it is interesting to speculate on why the French should more recently have put such emphasis on the interdependences between industrial sectors. It could well be that, as a relatively newly industrialised country, the external economies between industries which are essentially picked up in the concept of the *filière* were not present in France as they were in the more mature, industrialised countries of Britain and Germany.

The United Kingdom's tendency to excel in science but to fail to consummate early advantage is so well known that the label 'discovered in Britain but made elsewhere' now wears a little thin. A good part of the blame must surely fall on the élitist British education system which gives status to the arts and pure sciences over the applied sciences. This is exacerbated by the disproportionate share of the R&D budget going on defence and aerospace, offering the scientist the chance of highly sophisticated research at the leading edge of technology, even if the applications are limited. (France, too, has this problem.)

There has been a continuous schizophrenia in British industrial policy throughout the post-war period, with emphasis shifting regularly between intervention and competition (and not always in line with political swings). What began in the 1960s as an attempt to boost innovation was highjacked into a programme to promote national champions in science-based industries. This was subsequently dropped to be replaced after a while by a broader-based industrial strategy. Current policy puts emphasis on the free market, but with policy measures to promote awareness, finance (in particular to encourage the venture capital market) and, most recently, skills, it has distinct leanings towards the West German policies. But initiatives such as Alvey (in microelectronics) and Celltech (in biotechnology) have surprisingly interventionist (hare-like?) overtones.

The emphasis of German policy since the mid-1960s has been to encourage the take-up and application of new technologies with incentives to small firms, and general measures to encourage innovation and R&D. The Ministry for Research and Technology (BMFT) has run a series of specific programmes alongside these general measures, where, again, the emphasis has been upon the application of new technologies: that on data processing dates back to 1967 and was broadened in the mid-1970s to include components. The biotechnology programme dates from 1972, *before* the seminal developments in genetic engineering. In the event, in spite of these programmes, German industry has twice found itself effectively by-passed by developments in leading-edge technologies—in the early 1970s by developments in microelectronics, and in the mid-1970s by genetic engineering. It is interesting to note that in spite of the emphasis on small firms the bulk of the funding from such programmes went in fact to the largest firms in each sector. One of the dangers with the type of programme administered by the BMFT (and now by the Department of Trade and Industry in the United Kingdom) is the incestuous nature of advice received from industry on industrial priorities—there is a tendency to back the

Conclusions 287

favoured projects of the large firms. Although not as explicitly corporatist as the Japanese new technology programmes, it is sufficiently so to make simultaneous attempts to boost the small firm sector somewhat incongruous.

West Germany's greatest asset is its skill training programme which provides a work-force and middle management ready and able to learn about and make good use of new ideas. Time and again, when it comes to *applying* new technologies, it is those countries with the German traditions in skill training—West Germany, Sweden, Switzerland, and to a lesser extent the Netherlands—that come to the fore. In the sector studies it is a phenomenon observed in CAD and machine tools, in the remarkable turnaround in Siemens telecommunications capabilities after the EWS-A fiasco, and in the insistence on gateway architecture with the Prestel software. (What is the use of a system that cannot plug into data sets belonging to organisations such as banks?) It is currently observable in biotechnology, as West Germany's comparative strength in pharmaceuticals and chemicals begins to reassert itself. It reinforces the observation made earlier in relation to the increasing predominance of science-based industries that a necessary part of the infrastructure is a well-educated and well-trained workforce.

What of the other countries covered in these studies? The most noticeable feature is the relative strength of the Scandinavian presence in new activities. Thus, for both CAD, robotics and CNC-machine tools, Sweden ranks highest in Europe in terms of apparent usage (measured by relative market penetration). In biotechnology, firms such as Denmark's Novo Industry and Sweden's Pharmacia have established a world-wide reputation and are generally regarded as being front runners in their specialist areas (and incidentally are the only European firms quoted in the American stock market's biotechnology index). In telecommunications, Ericsson introduced its digital electronic exchange ahead of any of its European competitors and has been forging ahead in Third World (i.e. unprotected!) markets: in electronics, firms like Norsk Data are increasingly recognised as making a significant contribution to the industry. In offshore, after a slightly slow start, the Norwegians are now competing successfully with the major American firms in mainstream areas like rig building and—albeit behind a wall of protection—are building a sizeable presence in this industry.

The success of the Scandinavians is echoed to a lesser degree by the success of individual companies in some of the smaller European countries. Philips, for example, although it has had its vicissitudes, has held its own in consumer electronics while companies like AEG and Gründig have been in trouble. Heerema, again a Dutch company, has built itself into the world leader in heavy lifting gear for offshore developments. Olivetti is a leading force in world markets for automated machine tools and office equipment.

Perhaps the most disappointing performance has been that of Italy, which many had predicted in the early 1970s would be the leading European force in technological areas. The only area where it has forged ahead has been in automated machine tools, where a strong trade association has co-ordinated industry efforts. In other areas, as Dang Nguyen illustrates clearly in the case of telecommunications, political instability has had inevitable effects on industrial performance.

Europe's capacity to compete

What general lessons, if any, can be drawn from these 'themes' about Europe's capacity to compete in new industrial activities? At the beginning of this chapter we called each case study a 'snapshot' of the process of change and warned against generalisation. Certainly, one conclusion that can be drawn relates to the heterogeneity of experience, from sector to sector and from country to country. It could be argued that it is impossible to lay down any general policy guidelines because the only way in which policy can be approached is on a case-by-case, sector-by-sector approach. Although such a view has a certain appeal to those who feel that the right approach to the study of industrial change is from the grassroots upwards (a preference shared by all the authors of these studies), it nevertheless would appear unduly agnostic. The unifying focus on *how* new industrial activities are emerging and the commonality of experience revealed by the common themes explored earlier in this chapter make it worth trying to pull the threads together to present some kind of an overall perspective and some general policy conclusions.

THE HISTORICAL PERSPECTIVE

First, it seems worth trying to put the studies within an historical perspective to provide a broader overview of what seems currently to be happening to the European economies.

The general picture that has emerged of continuous evolutionary change, broken from time to time by important discontinuities which in their turn send ripples through the rest of the economy, accords well with an historical perspective of post-war development. According to such a view, the long boom of the 1950s and 1960s was based upon the exploitation of three main 'clusters' of technology: petrochemicals, the internal combustion and jet engine, and radio and TV electronics. The roots of these technologies went back well before the 1939–45 War, but they came into their own in the period of reconstruction and development after the War. This period of fast growth coincided with the time when these technologies were 'spilling over' in their secondary innovation phase: this was the period when, for example, we saw the introduction of many new man-made fibres and all kinds of new uses for plastics. Combined with Keynesian demand management techniques, this burgeoning of new products and processes led to an era of unprecedented and sustained growth throughout the world economy.

This era of growth and affluence came to a surprisingly sudden halt with the energy crises of the 1970s. Symptomatically, the world had exhausted immediate supplies of cheap energy on which the first two technological clusters were based; but the slowdown was also marked by strong cost push pressures from labour and commodity markets, and the saturation of European markets for many of the basic consumer goods and durables which had fuelled much of the long boom. Together, these factors (and Government reaction to them) brought an hiatus to growth, a sharp and continuing rise in unemployment and two deep recessions which have evoked memories of the 1930s.

Just as the 1930s are now seen as an important period in which the industrialised world adjusted from the coal-based technologies of the nineteenth century to the new technologies based on oil and electricity, so the 1980s are increasingly viewed as a period in which the world economies are adjusting to the new technologies of the end of the century. Chief amongst these is, of course, microelectronics, and its repercussions throughout the economy are chronicled in these six sector studies. Indeed, what we are currently witnessing is the beginning of the important secondary round of innovation and investment as microelectronics breaks out of its mainstream line of development (computers) and spills over into a host of new applications (for example CAD, videotex) and new process technologies (robotics, CNC machine tools, electronic telecommunications switching, etc.). And just as oil, energy and electronics provided a clustering of new technologies which became in many senses mutually reinforcing, so today we observe the same mutuality beginning to appear between microelectronics, new matierlas technologies and biotechnology.

Although Europe profited as much as any other group of countries from the period of fast growth in the 1950s and 1960s, technologically Europe has, throughout the post-war period, played second fiddle to the United States. For example, much of the fast growth of productivity achieved by the continental European countries throughout the 1950s and 1960s has been attributed to their process of 'catching up' with American best practice techniques. (The same is, of course, true of Japan—but Japan started from a lower base and moved faster.) The United States has been a willing partner in this process: technology has been readily available for licence, but perhaps the most noteworthy feature in this process of technology transfer has been the joint export of capital and management via the multinationals. For countries such as Holland (and now Ireland), the multinationals have transformed an agricultural country into an industrial one: in Belgium they have transformed an ageing industrial base into a modern service economy: in Britain, West Germany and even France (who has been least welcoming), they have played a vital role in the rebuilding and subsequent growth of the post-war economies. Without this injection of capital, technology and management, Europe's post-war economic history would have been very different.

While Europe certainly benefited from this process of technology transfer, the relationship was in many respects one of partnership rather than dependence. In the first place, it could be argued that many of the main 'inventions' lying behind the new technologies of that era (i.e. seminal developments in petrochemicals, aerospace, nuclear power, electronics) were European, and only subsequently transferred to the United States because of Hitler and the Second World War (the exodus of German intellectuals, the need to put the best brains into the Allied—and later NATO—defence effort, and the forcible transfer of German defence technologies to the United States in the aftermath of war). Second, Europe continued to prove a fertile source of new ideas. The civilian use of nuclear power, the development of radar and radio telescopes, penicillin, the jet engine—in all these areas Europe was in the lead and it was the Americans who picked them up and showed the Europeans how to develop them into commercial products. This in itself was a cause for concern but, given the austerity of post-war Europe, it was perhaps understandable, as was the attraction of the well-endowed American research laboratories to some of Europe's best brains. Technology transfer in the post-war period was very much a two-way process.

TECHNOLOGY GAPS AND MANAGEMENT GAPS

This suggests that there are two aspects to what is called the 'technology gap'. One relates to science and technology proper, the other relates to the commercial application of these ideas and more specifically to business management and production engineering than to basic science. The 'technology gap' of the 1950s and 1960s really concerned the latter; the United States exported its business systems to Europe but much of the scientific thinking that underlay the dominant technologies of that era had been European. Likewise, through widespread use of licensing and joint ventures, the Japanese built their economic miracle by importing both foreign technology and foreign business know-how, albeit adapting the latter to their own cultural traditions and evolving what today is regarded as the Japanese management paradigm.[2]

The much proclaimed 'technology gap' of the 1980s also has its double aspect. The weight of American spending on basic and applied research in microelectronics (mostly for defence) and health is widely perceived to have carried it ahead of its competitors. But the concept of the 'leading edge' in scientific studies is misleading. As the biotechnology study illustrates, the 'leading edge' is not a narrow strip of frontier with an identifiable apex, but rather a broad frontier from which advances may be made at many different point. In fact Europe has strength in many places on this front. Even in microelectronics, where in general the United States and Japan have taken the lead, the European contribution to, for example, software developments (illustrated in these studies by the French and British mastery of highly complex 3-D CAD systems) makes the picture far more complex than suggested by the black-and-white language of the technology gap.

What is, however, underlined by these studies is that where Europe lags most noticeably is in the *commercialisation and use* of new technologies—in other words that, as before, the gap is one of management not technology *per se*. But, whereas in the 1950s and 1960s, the emphasis was on economies of scale and mass production, with Europe learning from the United States, today it is the Japanese management system that dominates, with its long-term horizons and emphasis on flexibility, quality control and the importance of upgrading human capital.

How can Europe improve its performance?

If the gap that Europe faces is a management gap rather than a technological gap, then the scope for government action is limited. Governments can influence managers' decisions; they cannot make them. The emphasis of policy therefore has to be upon action which national Governments or the Commission can take to create an environment which encourages innovative decision-making. Abstracting at a very broad level, there would appear to be five main policy conclusions to emerge from these studies.

(i) CONCENTRATE ON USE RATHER THAN MANUFACTURE

The key to Europe's capacity to compete lies in its *use* of new technologies. Governments should not bow to the pressures which they meet from many quarters to establish and protect across-the-board capabilities in all new technologies. Instead, they should learn from the experience of Sweden, the country in Europe that has most successfully adjusted to and assimilated new technologies, and buy in skills and expertise as required from the United States, Japan or other European countries. Economic and technological chauvinism is infectious, leading, as the telecommunications experience illustrates, to the creation of 'national champions' and the undue fragmentation of the market. Nor does this necessarily argue for the creation of a manufacturing capability at a European level. There is the obvious danger in such cases of creating 'European champions' in place of national champions, and being forced to protect them from international competition because their products are not internationally competitive. Joint ventures and licensing deals with American or Japanese firms may being access to skills that are not available in Europe, and are costly and difficult to nurture from scratch. Inward investment, although it may bring useful management experience, may also pre-empt scarce skills and resources and fail to effect much substantive technology transfer.

(ii) PROVIDE CONTINUOUS STIMULUS TO COMPETITION

New industrial activities in the main emerge from the body of existing firms and industries, and the most important constituent in this process is the push for continuous, incremental innovation which emerges as firms respond to market pressures and technological opportunity. It requires an environment in which there is continuous stimulus to the firm from existing competitors and new entrants; in which new firms can take off and flourish and prick the complacency of established operators, and in which there is an awareness of technological opportunities and a willingness to seize them as they open up. Governments can help create such an environment by pursuing strong competition policies and eschewing special pleading for protection, by encouraging a venture capital market and a lively small firms sector, and by helping to create mechanisms which link academic science with the business community.

(iii) DO NOT NECESSARILY ESCHEW ALL SUPPORT FOR SUNRISE INDUSTRIES BUT TAKE HEED OF UNCERTAINTY AND OTHER COSTS

Arguments for the support of infant industries retain some validity in a world where protection, either overt or covert, is widespread and where substantial entry barriers are commonplace. Whatever the emphasis on the competitive process, it is not easy, for example, for an offshore supplies industry to establish itself in the face of competition from established majors with their pre-existing networks of sub-contractors and in an industry where the premium on experience makes dynamic learning curve economies a real barrier to entry. As the discussion of the *filière* makes clear, there may be a period in the early development of a new technology when there are externalities between the development of the mainstream technology and its upstream and downstream applications. But the concept of a 'strategic' industry is a very nebulous one, and the history of Government intervention to 'pick winners' has not been happy. The fact is that Governments, like firms, face uncertainty but, unlike firms, are not in business to take risks and do not have institutions adapted to do so. As French experience shows, Governments can play the risk business: but the cost may be very high and the pay-off uncertain. The costs, moreover, may not be merely financial. Support for infant industries may conflict directly with the objective of maximising the take-up of new technologies by users. Progammes of industrial support inevitably involve such trade-offs, and in the end it is for Governments to decide where priorities lie.

(iv) SUPPORT ACADEMIC SCIENCE, R&D, EDUCATION AND TRAINING PROGRAMMES

To be able to assimilate and make good use of new science-based technologies, an economy needs a strong base of science and technology, and skill training which endows the work-force with high-level but flexible skills. Academic science has both an infrastructure and a training function. On the one hand it helps to create an environment which both understands and contributes to the development of new technologies; on the other it alerts management to new ideas and helps train a cadre of scientists and technicians who are capable of using and extending these ideas. Traditionally, Europe has been a major source of creativity in new scientific ideas. There is some evidence that European creativity is being eclipsed by Japan and the United States, which raises the question of whether support for the basic and applied sciences has kept pace with needs, and especially whether a combination of rising student numbers and economies in public expenditures have not together squeezed academic research in Europe to too residual a role within the university system. More particularly, Europe lacks the mechanisms for linking academic research to industry. A lively venture capital market and small firm sector can play a part here, as can programmes for pre-competitive research linking firms and universities along the lines of ESPRIT (the European Commission initiative on microelectronics) and Alvey (its British counterpart). Just as crucial to the satisfactory use and assimilation of new technologies are the skills, flexibility and general educational levels of the work-force. A recent study contrasting the competence of the British and German shop-floor worker and supervisor in a matched sample of firms in the metalworking trades (Daly, Hitchens, Wagner, 1985) vividly illustrates why Britain finds it difficult to make good use of new technologies, and the advantage gained by the German plants in the higher and continuing level of skill training expected of their workers. Yet even West Germany contrasts its skill training programmes unfavourably with those of the Japanese, whose system of lifetime employment encourages the continuous upgrading of skills, and the retraining and redeployment of the work-force within the company.

Governments need not, indeed should not, meet all the costs of these support programmes. But the basic public good element in both research and education justifies sizeable Government contributions in both cases, and Governments need to provide a framework in which industry is encouraged and cajoled to make proper provision.

(v) RECOGNISE THE IMPORTANCE OF DEMAND AS WELL AS SUPPLY-SIDE POLICIES

New industrial activities do not emerge where there is no market. The interaction between supply (technology push) and demand factors has been a feature of these studies. It has also been suggested that we are currently at the point where microelectronics is spilling over from its mainstream activity into a host of new activities and bringing in its wake a clustering of new innovation. To pick up and make use of these innovations requires a new round of re-investment and equipment as firms in many sectors update product and process techniques. There is a danger that the relatively more restrictionist stance of European Governments, particularly Germany and the United Kingdom, will lock them out of that cycle of reinvestment which is a necessary part of the process of economic renewal. Economic history points to two occasions in the past when this has happened: to France in the 1880s and to Britain in the 1920s. On both occasions, with hindsight, economic historians have judged macroeconomic policies to have been unduly restrictive. While the problems of cost-push inflation may be more endemic today than they were in the past, there is clearly a danger that the current restrictive economic stance of European Governments is inhibiting this process of industrial renewal and exacerbating the problem of unemployment.

The issue therefore becomes to a lesser extent whether Europe *can* compete: rather, it is whether Europe *will* compete. Much emphasis has been put to date on the structural rigidities of the European economies, particularly the rigidities of the labour market. These studies suggest that while these rigidities play some part in Europe's somewhat laggard response to new technologies, the mote may be as much within the eye of the industrialists and Governments who make these assertions, as within the markets and institutions which they so readily criticise.

Notes

1. The *Wall Street Journal* of 31 January 1984 carried a supplement on the relative penetration and use of new technologies in Europe which indicated that opinion amongst multinational businessmen wholly bears out this perception of Scandinavian competence.
2. This term is used by the authors of the recent MIT study on automobiles (see Altschuler *et al.* 1984), where the impact of Japanese management systems (with their emphasis on training, quality control and long-range strategic planning) has been most marked.

Bibliography

Altschuler, A. A., Ross, D., Womack, J. and Jones, D. T. (1984), *The Future of the Automobile: Global Crisis and Transformation*, London, George Allen and Unwin, Boston, MIT Press.

Barker, R. (1984), 'Winning through the Biotechnology Discontinuity', paper delivered to Biotech '84, On-line Conferences, London.

Daly, A., Hitchens, D. and Wagner, K. (1985), 'Productivity Machinery and Skills in a Sample of British and German Manufacturing Plants', *National Institute Economic Review*, February, pp. 48–61.

Dosi, G. (1983), 'Technological Paradigms and Technological Trajectories: the determinants and direction of technical change and the transformation of the economy', Chapter 7 in *Long Waves in the World Economy* (ed.) C. Freeman, London, Frances Pinter.

Ergas, H. (1984), 'Why do some countries innovate more than others?', Centre for European Policy Studies, CEPS Paper No. 5, Brussels.

Freeman, C., Clark, J. and Soete, L. (1982), *Unemployment and Technical Innovation: A Study of Long Waves and Economic Development*, London, Frances Pinter.

Kuznets, S. (1940), 'Schumpeter's Business Cycles', *American Economic Review*, 30, No. 2, June, pp. 257–71.

Landes, D. (1969), *The Unbound Prometheus*, Cambridge, Cambridge University Press.

Malsot, J. (1980), 'Filières et pouvoirs de domination dans le système productif', in 'Les filières industrielles', *Annales des Mines*, January.

Martin, B. R., Irvine, J. and Turner, R. E. (1984), 'The Writing on the Wall for British Science', *New Scientist*, November 1984.

Mensch, G. (1975), *Das Technologische Patt: Innovationen Uberwinden die Depression*, Frankfurt, Umschau. English edition (1979), *Stalemate in Technology: Innovation Overcomes the Depression*, New York, Balinger.

Mistral, J. (1980), 'Filières et competivité: en jeux de politique industrielles', *Annales des Mines*, Les filières Industrielles, January.

Mowery, D. and Rosenberg, N. (1979), 'The influence of market demand upon innovation: a critical review of some recent empirical studies', *Research Policy*, 8, pp. 102–53.

National Academy of Science (1983), Committee on Science, Technology and Public Policy of the National Academy of Science, the National Academy of Engineering and the National Institute of Medicine. *Frontiers of Science and Technology: A Selected Outlook*, New York, N. A. Freeman.

Nelson, R. R. and Winter, S. G. (1982). *An Evolutionary Theory of Economic Growth*, Cambridge, Mass., The Belknap Press, Harvard University.

Northcott, J. and Rogers, P. (1985), *Microelectronics in Industry: An International Comparison: Britain, Germany, France*, Policy Studies Institute and the Anglo-German Foundation, London.

Rosenberg, N. (1982), *Inside the Black Box: Technology and Economics*, Cambridge, Cambridge University Press.

Schumpeter, J. A. (1912), *Theorie der Wirtschaftlichen Entwicklung,* Leipzig, Duncker and Humbolt.
Schumpeter, J. A. (1939), *Business Cycles: A Theoretical, Historical and Statistical Analysis of the Capitalist Process* (2 vols.), New York, McGraw Hill.
Surrey, J. and Walker, W. (1981), *The European Power Plant Industry: Structural Responses to International Market Pressures,* Sussex European Papers No. 12, Sussex European Research Centre, University of Sussex.

Index

Abu Dhabi 213
advanced machine tools 3, 5–6, 46–86, 265–6
   CNC 46, 48, 54–9, 61–2, 64, 66
   definition 46, 48
   exports 50, 70–6
   government policies 62, 65, 67–8, 78–83
   imports 58, 76–8
   in EEC 68–76
   in Europe 61–78, 79–83
   in Japan 14, 46, 52, 53, 57, 58–9, 65–6
   in USA 46–8
   industry 49–55
   international co-operation 56, 66–7, 82–3
   investment in 62, 67, 68, 82
   political implications 65–6, 78–83
   research and development 54, 62–3, 78
   spread of 55–61
   technical quality 70–6
   technology 47–9
advanced machine tools *see also* NC programming; robotics
Advanced Production Systems *see* APS
AEG 96
Aerospatiale 21
affinity chromatography 196
AGC 193
Agencie de l'Informatique 22
Agricultural Genetics Corporation *see* AGC
Agricultural Research Council (UK) *see* ARC
Agricultural University, Wageningen 194
agriculture 168, 172–3, 178, 184, 185, 188–9, 191, 194
AIF (Germany) 63
Ajinomoto 180, 186
Aker 230, 234, 256
Akzo-Organon 194
Alcatel-Thomson 93
Alfa Laval 196
Algeria 213, 238
Allen Bradley 54, 64
Alvey programme 282, 294
Amada 64
American Micro 274
American Telephone and Telegraph Corporation *see* AT&T
Amperex 274
AMREP 237

antibodies 165–6
Antiope-Télétel *see* Télétel
AOIP 98
Applicon 16, 17, 18, 20, 21, 24, 29
APS 37
ARC (UK) 18, 190
architecture 16, 32, 33
Aristo 18, 24, 25
ASCII 154–5
ASEA 32, 53, 60
Assigraph 21
Association Française 22
Association of Industrial Research Groups (Germany) *see* AIF
Association of Mechanical and Electrical Industries (Sweden) 35
Assoreni 195
ASST 104–5, 118, 119
W. S. Atkins 30
Atlas 2 computer 30
AT&T 4, 6, 267
   ESS 5 system 109
   interactive videotex 155, 157
   Rotary process 90
   telecommunications 88–9, 92, 94, 105–6, 109, 111–12, 118–19, 122, 124
Augustsson 230
Austria
   advanced machine tools 76
Autokon 35, 36
automatic transmission 90
automobile industry 12
   advanced machine tools 52, 67
   CAD 19, 23, 27
   robotics 54
Auto-trol 16, 17, 18, 21, 24, 29
AXE system 104
Azienda Statale del Servizio Telefonico *see* ASST

Baker Perkins 28
Baker Tools 228
Ballast Nedem 242
banking services 115, 144, 146, 152
Barclay's Bank 120
BASF 186
Battelle 21
Bayer 185, 186, 202, 209
BBC 141

Beesley Report 120
Bel 189
Belgium
  advanced machine tools 71, 72, 74
  biotechnology 182
Belgium *see also* EEC; Europe
Bell Canada 108, 155
Bell Laboratories 88, 94
Bendix 64
Bézier's patches 20
Bigfon *see* Project Bigfon
Bildschirmtext *see* BTX
bioelectronics 170, 174, 178
Biogen 177, 179, 191, 197, 203
bioreactors 163, 180, 185, 186, 187
biotechnology 4, 7, 161-212, 268-9, 282
  applications 171-4
  cell culture 163, 180, 185
  cell fusion 164-6, 179, 181, 186
  definition 162-4
  developments in 162-6, 198
  INA 162-4, 190
  ethics 199
  future of 206-9
  government policies 175, 179-80, 185, 187-90, 192, 195, 197, 203-9
  health and safety 199, 208
  history 162, 190
  in Europe 181-95, 198-209
  in Japan 179-81, 182-3
  in Scandinavia 195-7
  in Switzerland 197-8
  in USA 4, 175-9, 182-3, 200, 203-4
  industry 166-74, 176-98, 201-3
  international co-operation 181, 198-9, 204
  investment in 175, 177-8, 180, 183-190, 192-4
  patents 182-3, 197, 199-200, 202, 208
  rDNA 164-6, 173, 175, 179, 180, 181, 190, 196
  research and development 176-6, 180-1, 183-4, 189, 190, 192, 200, 207
  university research 176-7, 185, 187, 188, 190, 197, 205
Biotechnology Institute (Germany) 186
Biozyme 192
BMFT (Germany) 25-6, 37, 113, 185, 186, 205, 286
BNOC 227
BO 120
Boehringer Mannheim 185
Boeing Corporation 12, 282
Bosch 23, 54
Bouygues 237
BP 191, 227, 237
Bridgeport Textron 67

Britain
  advanced machine tools 64-7, 73-4, 80-1
  biotechnology 182, 183, 184, 190-3, 205-6
  CAD 14, 27-32
  computer industry 27-8
  Department of Energy 227, 228
  DoI 28, 31, 32, 39, 43-4, 144, 228
  DTI 28, 30, 193, 221, 286
  interactive videotex 139, 141-5, 156
  Ministry of Defence 28, 102
  Ministry of Technology 27, 30, 39
  North Sea oil 214, 216, 220-1
  offshore supplies industry 213, 220-9, 246, 248-9, 257-8
  OSO 226-7
  robotics 64-7
  telecommunications 4, 89, 96, 98, 100-4, 119-22, 124
Britain *see also* EEC, Europe
British Broadcasting Corporation *see* BBC
British National Oil Corporation *see* BNOC
British Radio Equipment Manufacturers' Association 141
British Technology Group 36, 193
British Telecom 4, 6, 104, 105, 120-2, 128
  Beesley Report 120
  interactive videotex 141-3, 149, 156, 157-8, 267
  Prestel 119, 120, 142-5, 147, 155, 267
British Telecom *see also* Martlesham Research Centre, Post Office (UK)
British Underwater Engineering 227
Britoil 227
John Brown 221, 223
Brown Boveri 23
Brown and Root 213, 221, 223, 241, 254
BNS-Gervais-Danone 189, 203
BTX 145-9, 156, 158
Bundesministerium für Forschung und Technologie (Germany) *see* BMFT
Bunker Ramo 48
Bureau de Recherche Petroleum 238
business information 144

CA Planning 26
cable television 111, 112, 114, 117, 121, 123
Cable and Wireless 120
Cables de Lyon 96
CAD 3, 6, 10-45, 264-5, 282, 285
  applications 16
  automobile industry 12, 19, 23, 27
  defence applications 11-12, 14, 19, 20, 23, 27
  definition 10

300 Index

CAD (*cont.*)
  distributed systems 15
  government policies 12, 19–44
  hardware 11–19
  in Europe 11, 15, 19–32, 37–44
  in Scandinavia 32–7
  in USA 11–19, 39
  industry 15–44
  integrated systems 23
  international co-operation 36–44
  investment in 12, 26, 31–2, 34
  pipework 30, 43
  political implications 26–7, 31–2, 34, 37–44
  research and development 14, 25–6, 35
  service bureaux 33, 35, 36
  shared logic systems 15
  software 11, 12
  turnkey systems 13–17, 23, 28, 30, 31, 33, 35
  university research 22, 29, 35, 63
  use patterns 37–44
  see also CAD/CAM, CAM, electronics industry, microelectronics
CAD80 35
Cadbury Schweppes 191
CADC 27, 30, 32
CAD/CAM 12, 14, 17, 23, 26, 29, 34–5, 36
  and advanced machine tools 49
  scheme 31
  see also CAD, CAM
CADIS 25
CADMAT scheme 31
CAL 18
Calcomp 24
Calgene 173
Calma 16, 17, 18, 43
  in Europe 21, 23, 24, 29
CAM 49, *see also* CAD, CAD/CAM
Cambridge Interactive Systems *see* CIS
Cambridge University 29, 30, 190
CAM-X 25
Canada
  Department of Communications 155
  interactive videotex 139, 147, 155, 157
  telecommunications 94, 113
CAO 19–23
car industry *see* automobile industry
Carlsberg 196
cartography 23, 37
CASS 36
cathode ray rubes *see* CRTs
CATIA 21, 41
CBS 155, 157
CCETT 149
CDC 14, 21, 24
cell culture 163, 180, 185

cell fusion 164–6, 179, 181, 186
Celltech 192, 193
cellulose degradation 205
Central Institute of Industrial Research (Norway) *see* SI
Centre Commun d'Études de Télécommunications et de Télévision *see* CCETT
Centre pour L'Exploration des Océans *see* CNEXO
Centre National d'Études Télécommunications *see* CNET
Centre Nationale de Recherche Scientifique *see* CNRS
CEPM 238, 239, 240
CEPT 147
Cetus 177, 179
CFEM 221, 235
CFP 238
CGCT 96, 97, 106, 117
CGE 96, 106, 116, 117, 151
chemical engineering 163, 178, 187
chemical industry 169, 173–4, 178, 180, 181, 189, 192, 194, 196
Chrysler Corporation 12
Ciba-Geigy 197
Cie. Française d'Entreprises Metalliques *see* CFEM
Cie. Générale Industrielle pour les Développements Opérationels des Recherches Sous-marines *see* GC Doris
Cie. Maritime des Expertises *see* COMEX
CII-Honeywell Bull 22, 152, 156
CIM 40, 43, 265
Cincinnati Milacron 48, 53, 60
CIS 18, 19, 40, 41, 43
  in Europe 19, 23, 24, 28, 29, 30, 32
  in Scandinavia 34
Cisi 21
CIT-Alcatel 68, 96, 97, 98, 106, 121
Citroën 12, 19–20
civil engineering 16, 27, 32, 33, 241
cloning *see* genetic engineering
Club 403 144
CNC 46, 48, 54–9, 79–82
  In Britain 64, 66
  in Germany 61–2, 80
  in Italy 64
CNET 106, 109
CNEXO 236
CNRS 20–1, 22, 188, 239
Comau.60
Comau-Fiat *see* Fiat
Combustion Engineering 228, 254
COMEX 221, 236, 240, 256
Comité d'Étude Pétroleum Marine *see* CEPM

Index

Committee for the Promotion of Life Sciences (Japan) 180
communication satellites 90, 99, 111, 115
  SBS 115
  TDF 1 124
  Telecom 1 115, 116
Compagnie Générale de Constructions Téléphoniques 96
Compagnie Générale d'Eléctricité *see* CGE
Company for Industrial Projects (Netherlands) 244
Compeda 18, 29, 30, 32, 44
computas 36
computer aided design *see* CAD
Computer Aided Design Centre *see* CADC
computer aided manufacture *see* CAM
computer industry 13–19, 28, 31–2
  interactive videotex 135, 138, 156
computer numerical control *see* CNC
computer-integrated manufacturing *see* CIM
computers 26, 34
  CIM 40
  hardware 11–19
  home computers 144
  in telecommunications 99–100
  industry 13–19, 28, 33–2, 135, 138, 156
  integrated systems 23
  interactive videotex 134
  mainframes 13, 31
  memory 13
  minicomputers 6, 13–15, 22–3, 26, 28, 36
  response time 13
  software 11
  turnkey systems 13–17, 23, 28, 30, 31, 33, 40
  virtual memory 13
Computers and Electronics Commission (Sweden) *see* DEK
Computervision 15, 16, 17, 18, 19
  in Europe 20, 21, 24, 25, 28, 29, 30, 32
  in Scandinavia 34, 35, 36
conception à l'aide d'ordinateur *see* CAO
concrete drilling platforms 230, 233, 236, 240, 241
Conférence Européenne des Postes et Télécommuncations *see* CEPT
Conflexip 236, 239
Consiglio Nationale delle Richerche 195
construction industry 23
Continental Shelf Act 214
Convention on the Community Patent 199
Corning Glass 128
Crédit Commercial de France 152
Creusot-Loire Enterprises 237
critical mass concept 200
Crossbar system 90, 102, 103, 108
CRTs 11, 13

Dalgety-Spillers 191
DAO 19, 22–3
Dassault 21
Data General 15
Datasaab 34
Datavision 21
DBP 90, 98, 102, 104, 112–14, 124, 128
  interactive videotex 139, 142–3, 145–9, 156
De Vlieg 66
DEC 14, 15
DECHEMA 185, 186, 204–5
defence industries 11–12, 14, 19, 20, 23, 27, 37, 81, 290
Degussa 185, 186
DEX 34–5
DEN Norske Veritas 36
Denmark
  biotechnology 182, 195–6
  see also Scandinavia
Department of Communications (Canada) 155
Department of Defense (USA) 12, 48
Department of Energy (UK) 227, 228
Department of Energy (USA) 175
Department of Industry (UK) *see* DoI
Department of Trade and Industry (UK) *see* DTI
dessin à l'aide d'ordinateur *see* DAO
Deutsche Bundespost *see* DBP
DFG 185
DGT 6, 98, 106, 114–17, 124, 128, 129
  interactive videotex 139, 147, 149–54, 156–7
DoI 28, 31, 32, 39, 43–4, 144, 228
Dietz 24, 25
Digital Electronics *see* DEC
digital switching *see* electronic switching
Direction Générale des Télécommunications *see* DGT
discontinuities 273, 275
Distillers Company 191
distributed systems 15
DMS system 104
DNA 162–4, 190
  rDNA 164–6, 173, 175, 179, 180, 181, 194, 196
double helix *see* rDNA
Douglas Aircraft Corporation 13, 282
Dow Jones 155
drilling equipment 215, 226, 228, 235, 239, 241
drilling platforms 215, 217–18, 221, 223, 235, 239, 241
  concrete 230, 233, 236, 240, 241
  jack-up rigs 241
drillships 221, 239, 240, 241

drug industry *see* pharmaceutical industry
DTI 28, 30, 193, 221, 286
Du Pont 178, 179, 186
Dynamit Novel 185
Dyvi 231

E10 system 127
EAX system 105
EEC
    advanced machine tools 57, 58, 68–70
    biotechnology 195, 198–203, 207
    CAD 38, 41, 44, 46–7
    foreign trade 68–76
    imports 76–8
    robotics 60
    telecommunications 92, 96, 127–8
EEC *see also* individual member states
Eidgenoessische Technische Hochschule 197
Eimac 274
EIO exchange 106
Ekofisk field 221, 229, 233, 236, 240
ELDAK 36
Electrolux 32, 53, 60
electronic directories 115, 150, 152
electronic mail 110, 144, 155
    see also interactive videotex
electronic money 115, 144
electronic news 151, 155
electronic switching 88, 90, 99–109, 127–8
electronics industry 14
    advanced machine tools 46, 49, 53–4
    bioelectronics 170, 174
    CAD 19, 22, 23, 36, 37
    interactive videotex 135–41
    telecommunications 99–109
electronics industry *see also* CAD, microelectronics, telecommunications
Elf Bio-Industries 188
Elf Bioresearch 188
Elf-Aquitaine 188, 205, 238, 239
Eli-Lilley 172, 179, 195
Ellerntel 94
Elliott Automation 27, 28
employment 232, 250–1
energy supplies 169, 170, 174, 180
ENI Group 195
Entreprise pour les Travaux Pétroleum Maritime *see* ETPM
enzyme immobilisation 182
enzyme technology 163, 173, 181, 185–7, 190, 192, 194–5
Ericsson 287
    AXE system 104
    CAD 32
    telecommunications 91, 93–4, 96–8, 105–8

H. Ernault Somua *see* HES
ESPRIT programme 44, 282, 294
ESS 5 system 109
Esso 241, 243–4, 245
Ethernet 128
ETPM 221, 236, 256
EUCLID 20
Eurolysine 189
Euronet/Diane 128
Europe
    advanced machine tools 46–7, 48–9, 61–78, 79–83
    biotechnology 4
    CAD 11, 15, 19–32, 37–44
    new technologies 5, 8–9, 263–95
    offshore supplies industry 7, 220–9, 235–59
    telecommunications 4, 89, 92–109, 111–22, 127–30
Europe *see also* EEC, individual states
European co-operation 11, 36–44
European Economic Community *see* EEC
Evans and Sutherland 29, 32
EWS-A system 102–4, 113, 114
EWS-D system 104, 127
exchange rates xv
exports *see* foreign trade

Face 98, 118
facsimile transmission 110, 115, 116
Fairchild 236, 274
Fanuc 53, 54
Farmitalia-Carlo Erba 195
FAST 198, 200
Fatme 98
FDA 164
Federal Republic of Germany *see* Germany
Federation of Scientific and Technical Associations (Italy) 194
fermentation processes 162–3, 179, 181–2, 184, 188–91, 194, 196, 203
Fermentation Research Institute (Japan) 180
Ferranti
    advanced machine tools 48
    CAD 18, 19, 25, 27, 28, 29
    offshore supplies industry 223
    telecommunications 121, 128
Fiat 12, 53, 60, 64, 98, 195
fibre optics 113–15, 117, 120–1, 124, 128
filière 116, 135–41, 148–9, 152–3, 156–7, 237, 244, 257, 267, 277–9
    definition 135–6
financing *see* investment
Flexible Manufacturing Systems *see* FMS
Flexservice 236, 239
FMS 14, 49

Food and Drug Administration (USA) see FDA
food industry 168, 172-3, 180-1, 185-6, 188-9, 191, 194, 199, 208
Foramer 235
Ford Motor Company 12
Forecasting Assessment in Science and Technology see TAST
foreign trade
  advanced machine tools 50, 58, 61, 63-4, 67, 68-78
  oil 230
  telecommunications equipment 92-4
Forex 221, 235, 256
Forschungszentrum des Deutschen Schiffbaus 26
A. B. Fortia 196
France
  advanced machine tools 67-8, 74, 75, 81-2
  biotechnology 182, 183, 184, 187-90, 205
  CAD 19-23
  computer industry 22
  interactive videotex 139, 147, 149-54, 156-7
  offshore supplies industry 213, 221, 235-40, 246, 247, 248, 257-8
  robotics 60, 67-8
  telecommunications 89, 96, 98, 105-7, 114-19, 124
  see also EEC, Europe
Fraunhöfer Institutes 26
Frigg field 229, 235, 236, 238, 240
FTZ 98
fuels 169, 170, 174
Fuji Electric 53, 60
Fujitsu 14, 96

G3S Group 96
gateways 145-6, 148, 151
GBF 185
GC Doris 221, 227-8, 230, 236, 237, 239
GDP 38-9, 49, 229-30, 243
GE 274
GEC
  4080 computer 145
  interactive videotex 135, 145-6, 148
  telecommunications 93, 96, 97, 102-3, 121
GEC Marconi 223
GEC Semiconductors 156
GECO 234
Genetech 164, 176, 178, 179, 196, 197
General Electric 53, 60
General Motors 12, 53, 60
generic technology 275

Genesys 28
genetic engineering 161-2, 164-6, 180, 185-8, 190, 193, 197, 201
  molecular genetics 186
  plant genetics 175, 178, 190, 191, 194, 205
  seed genetics 197, 205
Genex 179
Geomatriske Produkt Modeller see GPM
Geophysical Company of Norway 234
Geoteam 234
Gerber 16, 17, 18, 24, 29
German Machine Tool Manufacturers' Association see NDW
German Science Foundation see DFG
Germany
  advanced machine tools 52, 54, 56-7, 61-3, 74
  biotechnology 182, 183, 184-7, 204-5
  see also EEC, Europe
Gesellschaft für Biotechnologische Forschung see GBF
GI 274
Gildmeister 53, 62
Gist Brocades 173, 189, 193-4, 195-6
Gixi 31
Glaxo 191
GMW 18
Gotawerk Arendal 223
government policies 12, 285-8
  advanced machine tools 62, 65, 67-8, 78-83
  biotechnology 175, 179-80, 185, 187-90, 192, 195, 197, 203-9
  CAD 12, 29-44
  offshore supplies industry 226-9, 231-2, 237-40, 242-50, 254-9
  telecommunications 105-7, 114-15, 119-20
GPM 37
Graffenstraden 68
Grand Metropolitan 191, 197, 203
Graphael 21
Grays Tools 228
Great Britain see Britain
deGroot 243
Grootint 243
gross domestic product see GDP
growth hormones 196
GTE 93, 97, 105, 121, 155
GTE Italia 98, 109
Gulf of Mexico 213
de Gusto Shipbuilders 241

Halliburton 254
Halske 90
Harris 121

304 Index

Harvard Medical School 176, 186
HBG 241, 242, 243, 258
health care 171-2, 208-9
health and safety 199, 208
Heerma 228, 234, 241, 243-5, 254, 256, 258, 270
Alfred Herbert 53, 65
HES 68
Hitachi 14, 53, 60, 96
Hoechst 185, 186, 188, 191, 202, 205, 209
Hoffman La Roche 197
Holland *see* Netherlands
Hollandse Beton Group *see* HBG
home computers 144
Homelink 144
Honeywell 22
Howard 236
John Howards 228
Huré 67, 68
E. H. Hutton 196
hybridomas *see* monoclonal antibodies
Hybritech 179

IBA (UK) 141
IBM 6, 12, 14, 16-18, 43
    CAD 20, 23, 24, 27, 29, 34
    CATIA 21
    interactive videotex 155, 156
    robotics 53, 60
    SBS satellite 115
    telecommunications 107, 113, 119, 122, 126
IBM-Deutchland 146, 148
ICAM 12
ICAN 18, 36
ICI 191, 202
ICL 27, 28, 30, 31
IDA 121
IDI 25, 68
IFP 187, 188, 238, 239
IHC 241, 242
IKO Software Service 26
IMEG 221
imports *see* foreign trade
INCO 197
Independent Broadcasting Authority (UK) *see* IBA (UK)
industrial development 263-95
Industrial Reorganisation Corporation (UK) 65
industrial robots *see* robotics
Industrial and Technical Research Company (Norway) *see* SINTEF
Industry Act, 1972 (UK) 227
Information Displays Inc. *see* IDI

information providers 135, 138, 140, 142, 147, 156-7
information technology *see* interactive videotex
Informatique et Société Premier Plan 22
Ingersoll 65
INRA 187, 188
INSERM 187, 188
INSIS 128
Institut Français du Pétrole *see* IFP
Institut Pasteur 187, 188
Institute of Cell Research (Sweden) 196
insulin 195
Integrated Computer Aided Manufacturing *see* ICAM
Integrated Digital Access *see* IDA
Integrated Service Digital Network *see* ISDN
integrated systems 23
Intel 274
Intelautomatisme 68
Interactive Antiope *see* Antiope-Télétel
interactive graphics CAD *see* CAD
interactive videotex 134-60, 267-8, 272
    BTX 145-9, 156
    Club 403 144
    costs 142, 147, 152
    electronic directories 115, 150, 152
    electronic mail 144, 155
    electronic news 151, 155
    filière 116, 135-41, 148-9, 152-3, 156-7
    future of 154-8
    Home link 144
    in Canada 139, 147, 155, 157
    in Europe 141-54, 156-8
    in USA 154-5
    industry 135-68
    information providers 135, 138, 140
    Micronet 800 144
    Prestel 119, 120, 142-5, 147, 155
    private systems 135
    see also electronic mail, videotex
interferon 197
Intergraph 16, 17, 18, 24, 29
international co-operation 292-5
    advanced machine tools 56, 66-7, 82-3
    biotechnology 181, 198-9, 204
    CAD 36-44
    offshore supplies 219-20, 248
    telecommunications 108, 109
International Management and Engineering Group of Great Britain *see* IMEG
International Radio and Television Exhibition 145, 149
International Telephone and Telegraph Corporation *see* ITT

International Western Electric 89
Internationale de Service Industrielle et Scientifique *see* ISIS
investment 7, 63
   advanced machine tools 62, 67, 68, 82
   biotechnology 175, 177-8, 180, 183-9, 190, 192-4
   CAD 12, 26, 31-2, 34
   offshore supplies industry 228, 230, 232, 240, 243-4
   telecommuniations 106, 113-14, 115, 116-17, 118
   venture capital 8, 36, 187, 192, 193, 208, 282
IRI 96, 105
ISDN 6, 99, 110, 111, 112, 123, 125-7
   in Europe 114, 117
ISIS 239
islands of automation 10, 34, 43
Isopipe 30
Italpac 118
Italtel 93, 96-8, 104-5, 109, 118
Italy
   advanced machine tools 63-4
   biotechnology 182, 183, 184, 194-5
   offshore supplies industry 213
   robotics 53, 63-4
   telecommunications 89, 96, 97, 98, 104-5, 124
   see also EEC, Europe
ITT 88-9, 91, 93-4, 96-8, 105-7, 113, 118
   System 12 94, 102, 104

Jacobsson and Widmark 32
jack-up rigs 241
Japan 5
   advanced machine tools 14, 46, 52, 53, 57-9, 62, 79-80
   biotechnology 179-81, 182-3, 185-6, 203-4
   CAD 14-15
   foreign trade 76-8
   interactive videotex 155
   international co-operation 66-7
   Ministry of Agriculture 180
   Ministry of Health 180
   MITI 180
   robotics 14, 52, 53, 59-61
   STA 179-80
   telecommunications 96
jelly-industries 162

KabiGen 196
KabiVitrum 196
Kawasaki Heavy Industries 53, 60
Kearney 67

Kinshorne platform yard 228, 236
Kobe Steel 53, 60
Kockums Shipyards 32
Kommission für den Ansban technischen Kommunikationssgstene *see* KtK
Kondratiev long cycle 276
Kongsberg 18, 19, 24, 36, 234
Kornsnas Marma 196
Krone 113
KtK 113
Kuhlmann 25
Kuka 53, 60
Kvaerner 234, 256
Kyowa Hakko 180

Lacq field 238
Lafarge Coppée 189
Lansdale Tube 274
large-scale integration *see* LSI
Lasaffre 189
lead firms 237, 254-6, 259
Leicester University Biocentre 191
Lepetit 195
Lewis Offshore 228, 234
lifting equipment 218, 241, 270
Lignes Téléphoniques et Télégraphiques *see* LTT
LINE Group 67
LMT 94, 96
Lockheed Aircraft Corporation 12
locomotive technology 275-6
LSI 12, 16
LTT 94, 96
Luleaa Technical University 35
Lundy 29

McAuto Unigraphics *see* Unigraphics
McDermott 213, 221, 223, 254
McDonnell Aircraft Corporation 12, 13, 282
Machines Bull 21
Machines Françaises Lourdes Group *see* MFL
macrofilière see filière
mainframe computers 13, 31
Man Tech Program (USA) 48
management gaps 291
Mandell 64
manpower 27, 32, 66, 79
Manufacturing Technology Program (USA) *see* Man Tech Program (USA)
MAP 31
Marathon Le Tourneau 27, 235
Marconi 27, 28
Martlesham Research Centre (UK) 98, 103, 134, 141
Massachusetts General Hospital 186

306 Index

Massachusetts Institute of Technology *see* MIT
La Materiel Téléphonique *see* LMT
Matra
    CAD 18, 19, 21
    interactive videotex 151, 156
    telecommunications 96, 116, 117
Max Planck Institute *see* MPI
MCI 124
mechanical design 24–5
Medical Research Council (UK) *see* MRC
MEDUSA 23, 25, 40, 43
Mercury 120, 121, 122, 124
Messerschmidt 23, 185
MFL 67, 68
MICADO 22
microbial techniques 185
microelectronics 3, 10, 12, 31, 274–7
microelectronics *see also* CAD, electronics industry
microfilière see filière
Micronet 800 144
Microprocessor Application Project *see* MAP
microprocessors 87
military applications *see* defence industries
mineral ore working 174
minicomputers 6, 13–15, 22–3, 26, 36
    PDF–11 28
Ministry of Agriculture (Japan) 180
Ministry of Defence (UK) 28, 102
Ministry of Health (Japan) 180
Ministry of International Trade and Industry (Japan) *see* MITI
Ministry of Technology (UK) 27, 30, 39
Minitel 116, 117, 150–4, 158
Mission pour la Conception et le Dessin Assisté par Ordinateur *see* MICADO
MIT 14, 47–8, 186, 282
Mitel 113, 121
MITI (Japan) 180
Mitsubishi 67, 155
Mitsubishi Chemicals 180
Mitsui 155, 180
Mobil 235
mobile telephones 111, 114, 121
Mobilisation Plan for biotechnology (France) 189
modems 110
modular construction 218, 221, 223
Moët-Hennessy 189
molecular genetics 186
monoclonal antibodies 165–6, 181, 191, 197
monopolies 6
    interactive videotex 154–8
    telecommunications 89, 90–1, 98, 102, 104, 114, 117

Monsanto 173, 178, 179, 186, 197
Montedison Group 195
Mori Seika 67
Motorola 274
MPI 184–5, 186
MRC (UK) 183, 192, 193
MT20 system 127
multiplexing 90
Mupid 147

NAM 241, 243–4, 245
Nash equilibrium 107
National Enterprise Board (UK) *see* NEB
National Institutes of Health (USA) 175
National Research Development Corporation (UK) *see* NRDC
National Science Foundation (USA) 175
National Semiconductors 274
National Supply Co. 228
National Video 274
NBL enzymes 192
NC programming 10, 11, 19–20, 25, 32, 43, 46
    for metal processing 47–8
NC programming *see also* advanced machine tools, CNC, robotics
NDW 63
NEB 65, 227
NEC 93, 96
Nedlloyd 242, 243
Nestler 25
Netherlands
    biotechnology 182, 183, 194, 193–4
    offshore supplies industry 240–5, 213, 221, 246, 258
    see also EEC, Europe
network provision 109–22
neurobiology 197
new biotechnology *see* biotechnology
new industrial activity 79–83
new technologies
    advanced machine tools 3, 5–6, 46–86, 265–6
    background 1–2, 3–4, 6–9
    biotechnology 4, 7, 161–212, 268–9
    CAD 3, 6, 10–45, 264–5, 282, 285
    definition 2, 4–5
    developments in 4–5, 8–9
    future of 271–4, 282–5, 292–5
    government policies 4, 5, 285–8, 294–5
    history of 288–91
    interactive videotex 3, 6, 110, 115, 118, 134–60, 267–8, 272
    management gap 291
    microelectronics 3, 10, 12, 31, 274–7
    minicomputers 6, 13–15, 22–3, 26, 36

Index   307

offshore supplies industry 7, 213-62, 269-70, 277-8
  research and development 7, 8, 279-82
  risks 282-5
  robotics 14, 26, 32, 34, 35, 47-9, 51-5, 265-6
  technology gap 291
  telecommunications 3, 4, 5-6, 83-133, 266-7
  types of 3-4
Nippon Electric *see* NEC
Nixsdorf 113, 114
Nora-Minc report 114-15
Nordisk Fund 37
Norsk Data 36, 287
Det Norske Veritas 234
North sea oil 214, 216, 220-1, 222, 229-30, 238, 240-2, 253-4, 255, 270
Northern Telecom 93, 94-5, 105, 107, 108
  DMS system 104
Norway
  CAD 35-7
  offshore supplies industry 221, 229-34, 246, 248, 258
  see also Scandinavia
Norwegian Technology Research Council *see* NTNF
Nottingham Building Society 144
Novo Industry 173, 192, 195-6, 202, 287
NRDC (UK) 29
NTH/SINTEF 37
NTNF 35, 36
numerical control *see* advanced machine tools, NC programming
numerically controlled tools *see* advanced machine tools, NC programming

Oceaneering 236
OETB (UK) 228
Offshore Energy Technology Board (UK) *see* OETB
offshore oil industry 35, 36, 214, 221, 235, 238, 241
  North sea 214, 216, 220-1, 222, 229-30, 238, 240-2, 253-4, 255, 270
  offshore supplies industry 7, 213-62, 269-70, 277-8
    characteristics 215-19
    concrete drilling platforms 230, 233, 236, 240, 241
    costs 215-16, 218
    definition 213
    demands for 217
    development 217-19, 250-9
    drilling equipment 215, 226, 228, 235, 239, 241, 243

drilling platforms 215, 217-18, 221, 223, 235, 239, 241
drillships 221, 239, 240, 241
employment in 232, 250-1
filière 237, 244, 257
glossary 261-2
government policies 226-9, 231-2, 237-40, 242-50, 254-9
history 213-14, 235, 238, 241
in Europe 220-9, 235-59
in Scandinavia 221, 229-34, 246-9
in USA 7, 213, 221, 225, 228
international co-operation 219-20, 248
investment in 228, 230, 232, 240, 243-4
jack-up rigs 241
lead firms 237, 254-6, 259
lifting equipment 218, 241, 270
modular construction 218, 221, 223
North Sea 214, 216, 220-1, 222, 229-30, 238, 240-2, 253-4, 255, 270
piling techniques 218
problems 216
project management 261-17, 221, 223
research and development 218-19, 228-9, 233-4, 242, 247, 250-3, 256-7
rigs 235, 241
risks 216
ROVs 219, 249
surveying 215, 226
underwater work 215, 218, 219, 226, 236, 240, 243, 249
welding technology 218
Offshore Supplies Office (UK) *see* OSO
Oil Corporation of the Netherlands *see* NAM
oil recovery 174
oil rigs *see* drilling platforms
OKI 96
Okuma 66
Olivetti
  CAD 53, 54, 60, 64
  telecommunications 96, 98, 118
Olsen 230, 234
OPEC 174
open market 91, 227-8, 231, 248
optical fibres *see* fibre optics
Orsan 189
OSO (UK) 226-7, 229, 248
Oxford University 190

PABXs 94, 96, 98, 114, 121
PAM exchange 105
Paribas 188
patents
  biotechnology 182-3, 197, 199-200, 202, 208
PCBs 10, 14, 16, 20, 28, 30, 32, 42

308 Index

PDMS system 30, 41
PDP-11 minicomputer 28
Pechiney Ugine Kulhmann 188
Pelerin drillship 240
Pelican drillship 239, 240
Pentagone rig 235, 239, 240
Peugeot 67
pharmaceutical industry 166-7, 172, 178, 180-2, 188, 191, 195-6, 201-2
Pharmacia 196, 285
Pharmacia/Fortia 196
Philips 273
    CAD 54
    offshore supplies industry 244
    PRX system 109
    telecommunications 93-4, 96-7, 104, 109, 116
piling techniques 218
pipework 30, 43
Plan Calcul 21-2
Plan Composant 21-2
Plan Mini-Informatique 22
Plan Peri-Informatique 22
Plan Télématique 115, 149-50
plant genetics 175, 178, 190, 191, 194, 205
Plessey
    CAD 28
    offshore supplies industry 223
    telecommunications 93, 96, 97, 102-3, 121
political implications *see* government policies
Frederick Pollard 67
pollution 170, 174
Pontigga 64
Post Office (UK) 90, 98, 102, 119, 124, 134
    see also British Telecom
Pressed Steel Fisher 12
Prestel 119, 120, 142-5, 147, 155, 267
Prime Computer Company 14, 24, 30
printed circuit boards *see* PCBs
product cycle *see* filière
Project Bigfon 113
project management 216-17, 221, 223
Proteo system 105, 127
Proteo UT 10/3 105
Pruteen 191
PRX system 104, 109
PSS 120-1
PTTs 4
    interactive videotex 134-5, 138-40
    telecommunications 90, 91, 98, 101, 107-14, 122-30, 268
    see also individual authorities
public telephone authorities *see* PTTs
Pulse Amplitude Modulation *see* PAM
Pye 96, 102

Quante 113
Quest 18, 19, 28, 29, 32, 44

Racal 18, 44
    in Europe 19, 20, 21, 27, 28, 29, 32
Racal-Milicom 121
radar 11
radio transmission 90
Ramo-Casaneuve 67
Rank Hovis MacDougall 191
Ranland 274
Raytheon 274
RCA 273, 274
rDNA 164-6, 173, 175, 179, 180, 181, 194, 196
Readers Digest 155
Reading University 191
Rechnen und Entwicklungsinstitüt für EDV 26
recombinant DNA *see* rDNA
Recordati 195
remotely-operated sub-sea vehicles *see* ROVs
Renault 12, 20, 53, 60, 67, 68
research and development 7, 8, 279-82, 293-4
    advanced machine tools 54, 62-3, 78
    biotechnology 176-7, 180-1, 183-4, 189, 190, 194, 200, 207
    CAD 14, 25-6, 35
    neurobiology 197
    telecommunications 98, 102, 103, 108
    offshore supplies industry 218-19, 228-9, 233-4, 242, 247, 250-3, 256-7
response time 13
restriction enzymes 164
Reuters City Service 135, 144
Rhône Schelds Verole Corporation *see* RSV
Rhône-Poulenc 188
risk management 216, 282-5
robotics 265-6
    automobile industry 54
    CAD control 14, 26, 32, 34, 35
    definition 46, 48, 59
    development 47-9, 51-5, 80, 82
    industry 52-3
    Japanese 14, 52, 53, 59
    spread of 59-61
    see also advanced machine tools, NC programming
Rockwell 274
Rolls Royce 223
ROMULUS 29, 41
Roquette 189, 192
Rosenlev 53, 60
Roussel Uclaf 188

Index 309

ROVs 219, 249
Royal Bank of Scotland 144
Royal Boskalis-Westminster 241, 242, 243, 245
RSU 241, 243

SAAB-Scania 32, 35
safety *see* health and safety
Saga Petroleum 234
SAGE system 11
Saint Gobain 118
St. Mary's Hospital, London, 190
Saipen 256
Samprojetti 256
Sandoz 197
Sandvik 32
Sanofi 188
Saudi Arabia 94
SBS satellite 115
scaling up 216
Scandinavia
   biotechnology 195-7
   CAD 32-7
   new technologies 263-95
   offshore supplies industry 221, 229-34, 246-9
   telecommunications 89, 94, 108
   see also individual states
Scharmann 62
Schering AG 186
Schering Plough 179, 197
Schlumberger 235, 236, 254, 256
Schmidt International Fugro 243
Science and Engineering Research Council (UK) *see* SERC
Science Research Council (UK) *see* SERC
Science and Technology Agency (Japan) *see* STA
Scientific Council for Government Policy (Netherlands) 244
Sclavo 195
Scottish Development Agency 227
SCP 191, 202
SDL (AD 2000) 29
Secmai 18, 20, 21
Sedco 241
Sedneth 241
seed genetics 197, 205
SEFIS (UK) 65
SEL 23
   CAD 23
   interactive videotex 146, 148
   telecommunications 94, 96, 104, 113
SERC (UK) 192
service bureaux 33, 35, 36
sewage disposal *see* waste management
SGS-ATES 119

Shape Data 29, 30, 32, 41
shared logic systems 15
Shell 191, 193, 241, 243-4, 245
shipbuilding 19, 20, 26, 35, 37, 230, 231-2, 235-6, 241
SI (Norway) 35, 36, 37
SIAG 37
SI/Aker Group *see* SIAG
Siemens 273
   CAD 24, 25, 53
   EWS-A system 102-4
   EWS-D system 104
   interactive videotex 147
   Mupid 147
   telecommunications 90, 91, 93, 96, 97, 98, 107, 112, 114, 121
   teletex 113
SIGMA 21, 41
silicon valley 14, 36, 43
single cell proteins *see* SCP
SINTEF 37
SIP 104-5, 118-19
SIT-Siemans 96
Skyback 144
SLE-Citerel 106
small businesses 7, 144
   advanced machine tools 61
   CAD 16, 40, 43-4
   biotechnology 177, 187, 192
   robotics 52, 54-5
Small Engineering Firms Investment Scheme (UK) *see* SEFIS
SNA 119
SNEA 238
Societa Italiana per l'Esercizio Telefonico *see* SIP
Societa Produzione Antibiotici 195
SOCOTEL 98
Sona Biomedica 195
SOTELEC 98
Source 155
Soveno Institute (Italy) 195
Spain 149
specialised common carriers 111
Spinks Committee (UK) 192-3
STA (Japan) 179-80
Standard Electric Lorenz *see* SEL
Standard Radio and Telephone Company 34
Standard Telephone and Cables *see* STC
standardization 23
Stansaab Computer Company 34
Statfjord field 229, 232
Statoil 231, 234
STC 94, 96, 102-3
STET Group 96, 104-5, 118, 119, 128
Storm 21
Strowger process 90, 108

Index

STU (Sweden) 34, 35
Sumitomo 180
Summagraf 24
sunrise industries 2, 293
Surrey University 191
surveying 214, 218, 219, 226, 234
Sweden
    advanced machine tools 76
    biotechnology 182, 196
    CAD 32-5
    robotics 60
    telecommunications 89, 94, 108
    see also Scandinavia
Swedish Economic Development Agency 34
switching systems *see* electronic switching
Switzerland
    advanced machine tools 76, 77
    biotechnology 182, 197-8
SX/100 exchange 113
Sylvania 274
System 12 94, 102, 104
System X 102-4, 105, 118, 120, 121
SYSTRID 21, 31

Takeda Chemical Industries 180, 197
Takisawa 67
Takuba Science City 180
Tate and Lyle 191
TDF 1 satellite 124
Technip 237
Technology Development Directorate (Sweden) *see* STU
technology gaps 291
Technology Institute of Graz 147
technology transfer 193, 207
Technovision 25
Tekade 96, 97, 104
Tektronix 13, 14, 29
Telecom 1 satellite 115, 116, 124
telecommunications 3, 4, 5-6, 87-133, 266-7
    automatic transmission 90
    AXE system 104
    cable television 111, 112, 114, 119, 121, 123
    communication satellites 90, 99, 111, 115, 116, 124
    computers for 99-102
    Crossbar system 90, 102, 108
    developments in 109-22
    definition 87-8
    DMS system 104
    EAX system 105
    electronic mail 110
    electronic switching 88, 90, 99-102
    electronics for 99-109

    ELO exchange 106
    equipment 90
    ESS 5 system 109
    EWS-A system 102-4, 113, 114
    EWS-D system 104
    facsimile transmission 110, 115, 116
    future of 122-7
    government policies 105-7, 114-15, 119-22
    history of 89-99, 109-22
    in Europe 4, 89, 92-109, 111-22, 127-30
    in Scandinavia 89, 94, 108
    in USA 4, 6, 88-9, 93-4, 109, 111-12, 114, 122, 128-9
    industry 89-122, 127-31
    international co-operation 108, 109
    investment in 106, 113-14, 115, 116-17, 118
    ISDN 6, 99, 110, 111, 112, 114, 117, 123, 125-7
    mobile telephones 111, 114, 121
    modems 110
    multiplexing 90
    network provision 109-22
    PABXs 94, 96, 98, 114, 121
    PAM exchange 105
    Prestel 119
    Project Bigfon 113
    Proteo system 105
    Proteo UT 10/3 105
    Prx system 104, 109
    PTTs 4, 90, 91, 98, 101, 107-14, 122-30, 268
    radio transmission 90
    regulations 4, 111, 119-22, 126
    research and development 98, 102, 103, 108
    Rotary process 90
    specialised common carriers 111
    Strowger process 90, 108
    SX/100 exchange 113
    System 12 99, 102, 104
    System X 102-4, 119, 120, 121, 127
    teleconferencing 110-11
    telematics 87, 90, 99, 104, 112, 114-15, 123
    telephone networks 110, 111, 112, 123
    teletex 110, 113
    TXE exchanges 102
    videocommunications 110, 111, 123
    videotex 110, 115, 118, 134-60
    voice communication 87-90
    see also electronics industry, microelectronics
teleconferencing 110-11
Telefonbau Normalzeit *see* TUN

Index

telematics 87, 90, 99, 109, 112, 114–15, 123
Telenet 155
telephone networks 110, 111, 112, 123
    interactive videotex 140
    mobile telephones 111
    see also voice communication
Télépresse 149
Telerate 144
Télétel 115, 147, 149–54, 155, 157, 158
Télétel 3V 150, 151, 153
teletex 110, 113
Telettra 98, 105, 109
television 124, 157
    cable 111, 112, 114, 117, 121, 123
Telic 150–1, 156
Telidon 147, 155
Texas Instruments 53, 60, 274
Third World countries 91
Thomson 96, 97, 106–7, 117, 151, 273
    nationalization 116
Thorn 96, 102
3D design 16, 20, 21, 25, 291
TI Matrix Churchill 67
Times–Mirror 155
TNO 242
Total 238, 239
Toulouse University 188
Toyoda 68
Toyota 68
Trallfa 53, 60
Transgène 188
Transpac 115, 117, 149, 151–2, 156–7
travel reservations 143, 144
Trecker Martin 67
Trosvik 230
TRT 96, 116, 151, 156
Trumpf 62
TUN 96
turnkey systems 13–17, 23, 28, 30, 31, 33, 40
Trondheim Technical University 37
2D design 16, 20, 25
TXE exchanges 102
Tymnet 155

UCIMU 63
UIE 221, 235, 237, 256
Ulstein 230
uncertainty see risk management
underwater work 215, 218, 219, 226, 236, 240, 243, 249
Unigraphics 18, 24, 29
Unilever 191, 193, 194, 203, 244
Unimation 53, 60
Union Industrielle et d'Entreprise see UIE
UNISURF 20

United Kingdom see Britain
United States Air Force see USAF
United States of America see USA
University College, London 197
university research 4, 280, 293–4
    biotechnology 176–7, 185, 187, 188, 190, 197, 205
    CAD 12, 29, 35, 63
Upper Clyde Shipyard 227
Uppsala University 196
USA
    advanced machine tools 46–8, 58–9
    biotechnology 4, 175–9, 182–3, 200, 203–4
    CAD 6, 11–19, 39
    Department of Defense 12, 48
    Department of Energy 175
    FDA 164
    interactive videotex 154–5, 157
    National Institutes of Health 175
    National Science Foundation 175
    offshore supplies industry 7, 213, 221–5, 228
    telecommunications 4, 6, 88–9, 93–4, 109, 111–12, 114, 122, 128–9
USAF 12

VDMA 54
Velec 117
venture capital 8, 36, 187, 192, 193, 208, 282
Verband Deutscher Maschinen und Anlagenbau see VDMA
Verbrauchers Bank 146
Verein Deutscher Ingenieure 27
Very High Speed Integrated Circuit program see VHSIC
very large-scale integration see VLSI
Vetco 228
veterinary medicine 172
VHSIC 12
Vickers Oceanonics 221, 227, 249
video telephony see interactive videotex
videocommunications 110, 111, 123
videocommunications see also interactive videotex
videotex 3, 6, 110, 115, 118
    see also interactive videotex
Viewdata see interactive videotex, Prestel
Viewtron 155
virtual memory 13
VISTA 155
VLSI 12, 43
voice communication 87–90
voice communication see also telephone networks
Volker-Stevein 242

Volkswagen Foundation 185
Volkswagenwerk 52, 53, 59–60
Volvo 32, 196

Wagner Commission (Netherlands) 244
waste management 170, 171, 174, 185, 194, 202
welding technology 218
Wellcome Foundation 191
West Africa 213, 238, 239, 240
West Germany *see* Germany
Western Electric 88, 93, 94, 107
Western Union 88

Westinghouse 53, 60, 274
WF-RDM Groupsmaatschappij BV 241
Whitbread 191
Wilhelmsen 230
Wilton-Rotterdam Dockyard Group *see* WF-RDM Groupsmaatschappij BV
Wimpey 223
Wistar Institute (USA) 175

Xerox 128

Yamazaki 67
Yaskawa Electric 53, 60
Yasuda 67

THE OPEN UNIVERSITY

Ex-Libris   E. J. Tait
Systems Group Technology Faculty
The Open University
Walton Hall
Milton Keynes MK7 6AA
Telephone: Milton Keynes (0908) 653759